AJ Sadler

Mathematics
Applications

T0357905

Student Book

Unit 1

Mathematics Applications Unit 1
1st Edition
A.J. Sadler

Publishing editor: Robert Yen
Project editor: Alan Stewart
Cover design: Chris Starr (MakeWork)
Text designers: Sarah Anderson, Nicole Melbourne,
Danielle Maccarone
Permissions researcher: Jessica Boland
Answer checker: George Dimitriadis
Production controller: Erin Dowling
Typeset by: Cenveo Publisher Services

Any URLs contained in this publication were checked for currency
during the production process. Note, however, that the publisher
cannot vouch for the ongoing currency of URLs.

For product information and technology assistance,
in Australia call **1300 790 853**;
in New Zealand call **0800 449 725**

For permission to use material from this text or product, please email
aust.permissions@cengage.com

National Library of Australia Cataloguing-in-Publication Data
Sadler, A.J., author.
Mathematics applications : unit 1 / A.J. Sadler.

1st revised edition
9780170390194 (paperback)
Includes index.
For secondary school age.

Mathematics--Study and teaching (Secondary)--Western Australia.
Mathematics--Textbooks.

510.712

Cengage Learning Australia
Level 7, 80 Dorcas Street
South Melbourne, Victoria Australia 3205

Cengage Learning New Zealand
Unit 4B Rosedale Office Park
331 Rosedale Road, Albany, North Shore 0632, NZ

For learning solutions, visit **cengage.com.au**

Printed in Malaysia by Papercraft.
11 24

PREFACE

This text targets Unit One of the West Australian course *Mathematics Applications*, a course that is organised into four units altogether, the first two for year eleven and the last two for year twelve.

This West Australian course, *Mathematics Applications*, is based on the Australian Curriculum Senior Secondary course *General Mathematics*. With only very slight differences between the Unit One content of these two courses this text is also suitable for anyone following Unit One of the Australian Curriculum course, *General Mathematics*.

The book contains text, examples and exercises containing many carefully graded questions. A student who studies the appropriate text and relevant examples should make good progress with the exercise that follows.

The book commences with a section entitled **Preliminary work.** This section briefly outlines work of particular relevance to this unit that students should either already have some familiarity with from the mathematics studied in earlier years, or for which the brief outline included in the section may be sufficient to bring the understanding of the concept up to the necessary level.

As students progress through the book they will encounter questions involving this preliminary work in the **Miscellaneous Exercises** that feature at the end of each chapter. These miscellaneous exercises also include questions involving work from preceding chapters to encourage the continual revision needed throughout the unit.

Some chapters commence with, or contain, a **'Situation'** or two for students to consider, either

individually or as a group. In this way students are encouraged to think and discuss a situation, which they are able to tackle using their existing knowledge, but which acts as a fore-runner and stimulus for the ideas that follow. Students should be encouraged to discuss their solutions and answers to these situations and perhaps to present their method of solution to others. For this reason answers to these situations are generally not included in the book.

At times in this series of books I have found it appropriate to go a little beyond the confines of the syllabus for the unit involved. In this regard, when considering simple interest I include the concepts of daily balance and minimum monthly balance and, whilst the syllabus inclusion of 'composite shapes', allows the area of a trapezium to be considered, this text includes the trapezium as one of the basic shapes along with those specifically mentioned in the syllabus. Also in the chapters on perimeter, area and volume, I have included some 'inverse questions' which require the reader to determine a length given a perimeter, area or volume, thus going beyond the syllabus requirements of simply finding perimeter, area and volume. These inverse questions may require the use of some basic equation solving techniques. I leave it up to the readers and teachers to decide whether to cover these aspects or not.

Alan Sadler

ISBN 9780170390194

CNTENTS

IMPORTANT NOTE

This series of texts has been written based on my interpretation of the appropriate *Mathematics Applications* syllabus documents as they stand at the time of writing. It is likely that as time progresses some points of interpretation will become clarified and perhaps even some changes could be made to the original syllabus. I urge teachers of the *Mathematics Applications* course, and students following the course, to check with the appropriate curriculum authority to make themselves aware of the latest version of the syllabus current at the time they are studying the course.

Acknowledgements

As with all of my previous books I am again indebted to my wife, Rosemary, for her assistance, encouragement and help at every stage.

To my three beautiful daughters, Rosalyn, Jennifer and Donelle, thank you for the continued understanding you show when I am 'still doing sums' and for the love and belief you show in me.

Alan Sadler

ISBN 9780170390194

PRELIMINARY WORK

This book assumes that you are already familiar with a number of mathematical ideas from your mathematical studies in earlier years.

This section outlines the ideas which are of particular relevance to Unit One of the *Mathematics Applications* course and for which some familiarity will be assumed, or for which the brief explanation given here may be sufficient to bring your understanding of the concept up to the necessary level.

Read this 'Preliminary work' section and if anything is not familiar to you, and you don't understand the brief explanation given here, you may need to do some further reading to bring your understanding of those concepts up to an appropriate level for this unit. (If you do understand the work but feel somewhat 'rusty' with regards to applying the ideas some of the chapters afford further opportunities for revision, as do some of the questions in the miscellaneous exercises at the end of chapters.)

- Chapters in this book will continue some of the topics from this preliminary work by building on the assumed familiarity with the work

- The **miscellaneous exercises** that feature at the end of each chapter may include questions requiring an understanding of the topics briefly explained here.

Types of number

It is assumed that you are already familiar with:

Counting numbers	$1, 2, 3, 4, 5, 6, 7, \ldots$
Whole numbers	$0, 1, 2, 3, 4, 5, 6, 7, \ldots$
Integers	$\ldots, -5, -4, -3, -2, -1, 0, 1, 2, 3, 4, 5, 6, 7, \ldots$

It is also anticipated that you are familiar with *fractions* and *decimals*, including negatives, that you can add, subtract, multiply and divide such numbers (with a calculator when appropriate) and are able to convert between these forms of representation.

Powers

You should already be familiar with the idea of raising a number to some *power* (perhaps squared, cubed etc.), the idea of the *square root* or *cube root* of a number and be familiar with zero and negative integers as powers.

For example

$$5^2 = 5 \times 5 = 25$$

$$6^3 = 6 \times 6 \times 6 = 216$$

$$2^0 = 1$$

$$3^0 = 1$$

$$\sqrt{25} = 5$$

$$\sqrt[3]{216} = 6$$

$$2^{-3} = \frac{1}{2^3} = \frac{1}{8}$$

$$3^{-2} = \frac{1}{3^2} = \frac{1}{9}$$

You are probably also familiar with numbers expressed using *standard form* or *scientific notation*, e.g.

$$260\,000 = 2.6 \times 10^5,$$

$$53\,200\,000 = 5.32 \times 10^7,$$

$$0.000042 = 4.2 \times 10^{-5},$$

$$0.0015 = 1.5 \times 10^{-3}.$$

Rule of order

An ability to correctly apply the *rule of order* is assumed. For example, when evaluating $2 + 3^2$ you should know to square the 3 first and then add the 2:

$$2 + 3^2 = 2 + 9$$
$$= 11$$

This rule of order is sometimes remembered as BIMDAS:

Brackets

Indices

Multiplication and **D**ivision in the order they occur

Addition and **S**ubtraction in the order they occur

Rounding and truncating

Answers to calculations may need *rounding* to a suitable or specified accuracy. For example:

$193.3 \div 17 =$ 11.370 588 if we round our answer to six decimal places

11.370 59 if we round our answer to five decimal places

11.370 6 if we round our answer to four decimal places

11.371 if we round our answer to three decimal places

11.37 if we round our answer to two decimal places

11.4 if we round our answer to one decimal place

11 if we round our answer to the nearest integer.

In some cases the situation may make *truncating* more appropriate than rounding. Suppose for example we have $10 and wish to buy as many chocolate bars costing $2.15 each as possible. Whilst $10 ÷ $2.15 is 4.65 if we round to two decimal places, 4.7 if we round to one decimal place and 5 if we round to the nearest integer, a more appropriate answer is obtained by truncating to 4 as that is the number of chocolate bars we would be able to buy with our $10 (and we would have $1.40 change). If we truncate to an integer we discard the decimal part entirely.

It is assumed that you can round appropriately in order to obtain an approximate answer to a calculation (i.e. you can *estimate* the answer).

Ratios

The idea of comparing two or more quantities as a *ratio* should be something you are familiar with.

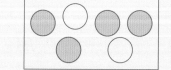

For example, in the diagram the ratio of white circles to blue circles is $\quad 2 : 4$,

which simplifies to $\quad 1 : 2$.

Suppose the ratio of males to females in a school is $17 : 21$. If we know that there are 231 females in the school we can determine the number of males.

Males : females $= \quad 17 : 21$ } $\times 11$

$= \quad ? : 231$

The number of males $= 17 \times 11$, i.e. 187.

231 / 21

11

Rates

A rate shows how one quantity changes with relation to another. For example, how the total number of items produced may increase for each extra hour of production, how the total cost changes with each extra kilogram we purchase, how the total distance we have travelled increases with each extra second we travel, etc.

Rate skills

Some other examples of rates used in everyday life are given below.

Cost of fuel: cents/litre	Cost of meat: $/kg	Postal rates: $/kg	Pay rate: $/hour
Pulse: beats/minute	Infection rate: cases/year	Download rate: bytes/second	Speed: km/hour
Typing rate: words/minute	Density: g/cm^3	Sports rates: points/game	Sports rates: runs/over

In its simplest form a rate is usually expressed as 'per 1 unit'.

Thus a rate of $60/2 kg would usually be written as $30/kg.

31 heart beats/half minute would usually be written as 62 beats/minute.

However rates are not always expressed in this simplest form. For example fuel consumption is sometimes given as litres/100 km and when rate involves the population of a country some rates are given as 'per 1000 of population', as in birth rates, death rates, etc.

A rate often involves a comparison between two quantities that have different units of measure. For example kilometres per hour, dollars per kilogram etc. However the word rate is also frequently used when two quantities involved in the rate have the same unit. The rate may then be given as a percentage rate. Some examples are given below.

Tax rate: e.g. 10% goods and services tax (GST), e.g. 30% income tax.	Rate of commission: e.g. $20 commission for each $500 of sales, i.e. 4% commission.
Conviction rate: e.g. 38 cases in each 100 result in a conviction, i.e. 38% conviction rate.	Unemployment rate: e.g. 58 unemployed people per 1000 possibles, i.e. 5.8% unemployment.
Annual death rate: e.g. 16 deaths per 1000 of population, i.e. 1.6% death rate.	Insurance rate: e.g. $3.50 per $1000 insured, i.e. 0.35% of value.

ISBN 9780170390194

Percentages

It is assumed that you know, or can determine, the common equivalences between decimals, fractions and *percentages*. Some of these are shown below:

Decimal	Fraction	Percentage
0.01	$\dfrac{1}{100}$	1%
0.1	$\dfrac{1}{10}$	10%
0.2	$\dfrac{2}{10}$ i.e. $\dfrac{1}{5}$	20%
0.25	$\dfrac{25}{100}$ i.e. $\dfrac{1}{4}$	25%
0.5	$\dfrac{5}{10}$ i.e. $\dfrac{1}{2}$	50%
0.75	$\dfrac{75}{100}$ i.e. $\dfrac{3}{4}$	75%
0.8	$\dfrac{8}{10}$ i.e. $\dfrac{4}{5}$	80%
1.1	$1\dfrac{1}{10}$	110%

Whilst it is also assumed that you would have previously encountered the ideas of *expressing some amount as a percentage of a total amount* and of *finding percentages of amounts* these skills will be revised and further practiced in the chapter on percentages.

Understanding formulae

The rule $C = 2 \times \pi \times r$ tells us how we can determine C, the circumference of a circle, knowing π and r, the radius of the circle. It is assumed that you are already familiar with the convention of leaving out the multiplication sign(s) and writing the formula simply as $C = 2\pi r$.

Similarly,

$A = b \times h$	is written	$A = bh$
$V = I \times R$	is written	$V = IR$
$v = u + a \times t$	is written	$v = u + at$
$A = \pi \times r \times r$	is written	$A = \pi r^2$

Perimeter

The perimeter of a closed shape is the distance all around the outside. It is the length of the outline of the shape. (For circles the perimeter is known as the *circumference*.)

For the shape shown the perimeter is 34 cm.

(= 10 cm + 5 cm + 4 cm + 5 cm + 3 cm + 2 cm + 5 cm)

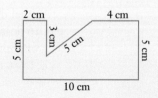

Area

Chapter 8 involves finding perimeters and areas of various shapes. The chapter assumes you are already familiar with determining the area of rectangles, triangles, parallelograms, trapeziums and circles and with determining the perimeters of such shapes, given sufficient information.

The formulae for the areas of these shapes, and for the circumference of a circle are given below by way of a reminder.

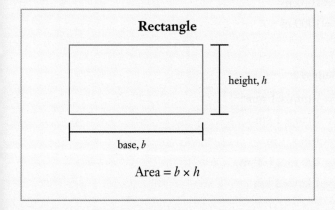

Rectangle

height, h

base, b

$$\text{Area} = b \times h$$

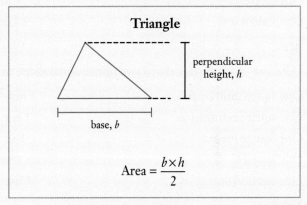

Triangle

perpendicular height, h

base, b

$$\text{Area} = \frac{b \times h}{2}$$

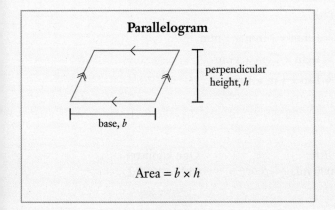

Parallelogram

perpendicular height, h

base, b

$$\text{Area} = b \times h$$

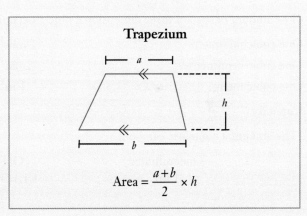

Trapezium

a

h

b

$$\text{Area} = \frac{a+b}{2} \times h$$

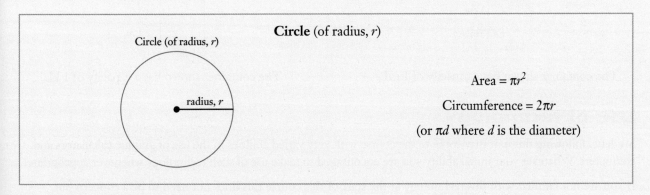

Circle (of radius, r)

Circle (of radius, r)

radius, r

$$\text{Area} = \pi r^2$$

$$\text{Circumference} = 2\pi r$$

(or πd where d is the diameter)

Units of length, area, volume and capacity

It is assumed that you are familiar with the following metric units.

Length

One millimetre:	1 mm
One centimetre:	1 cm = 10 mm
One metre:	1 m = 100 cm
One kilometre:	1 km = 1000 m

Area

(The size of the surface that a two dimensional shape occupies.)

One square millimetre:	$1 \text{ mm}^2 = 1 \text{ mm} \times 1 \text{ mm}$
One square centimetre:	$1 \text{ cm}^2 = 1 \text{ cm} \times 1 \text{ cm}$
One square metre:	$1 \text{ m}^2 = 1 \text{ m} \times 1 \text{ m}$
One hectare:	$1 \text{ ha} = 100 \text{ m} \times 100 \text{ m}$
One square kilometre:	$1 \text{ km}^2 = 1 \text{ km} \times 1 \text{ km}$

Volume

(The amount of space a three dimensional object occupies.)

One cubic millimetre:	$1 \text{ mm}^3 = 1 \text{ mm} \times 1 \text{ mm} \times 1 \text{ mm}$
One cubic centimetre:	$1 \text{ cm}^3 = 1 \text{ cm} \times 1 \text{ cm} \times 1 \text{ cm}$
One cubic metre:	$1 \text{ m}^3 = 1 \text{ m} \times 1 \text{ m} \times 1 \text{ m}$

Capacity

(The amount a container can hold.)

One millilitre:	One litre:	One kilolitre:
1 mL	1 L = 1000 mL	1 kL = 1000 L

The container shown has a capacity of 1 mL.   The container shown has a capacity of 1 kL.

Use of technology

Students following this unit may come to the course with very varied abilities in the use of graphic calculators and computers. Whatever your initial ability you are encouraged to make use of such technology whenever appropriate.

If you are not familiar with this technology at the start of this course this does not need to be a concern but, as the course progresses, do try to become familiar with entering data into the columns of a spreadsheet on a computer or calculator and with carrying out straightforward operations on those entries such as finding the total of a list of numbers etc.

ISBN 9780170390194

1.

Use of formulae

- Use of formulae
- Use of formulae in spreadsheets
- Miscellaneous exercise one

Situation One

A person has a modern cookbook that gives oven temperatures in degrees Celsius but an old oven with the dial graduated in degrees Fahrenheit.

The person wishes to convert some Celsius temperatures in the book to Fahrenheit temperatures in order to correctly set the oven dial.

A book she has tells her that to change a Celsius temperature (°C) to the equivalent Fahrenheit temperature (°F) she can use the following rule, or **formula** (plural: formulae or formulas):

$$F = \frac{9}{5}C + 32.$$

Change the following Celsius temperatures to Fahrenheit.

a　200°C

b　170°C

c　150°C

d　100°C

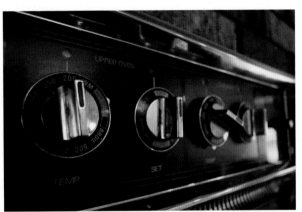

Situation Two

For some medical procedures, for example administration of chemotherapy, the surface area of the patient's body may need to be known. This is not an easy thing to measure so instead there are various formulae that can be used. One such formula, known as the Mosteller formula, provides an estimate of the surface area of the patient, A m^2, using the weight of the patient, W kg, and the height of the patient, h cm.

The Mosteller formula:
$$A = \sqrt{\frac{W \times h}{3600}}$$

According to this formula estimate the surface area of a patient with:

a　weight 80 kg, height 169 cm.

b　weight 61 kg, height 155 cm.

c　weight 79 kg, height 183 cm.

d　weight 70 kg, height 159 cm.

ISBN 9780170390194

Use of formulae

On the previous page, the formula

$$F = \frac{9}{5}C + 32$$

allowed values of F to be determined for given values of C, and the formula

$$A = \sqrt{\frac{W \times h}{3600}}$$

allowed values of A to be determined for given values of W and h.

The formulae on the previous page could have been stated in words rather than symbols. For example $F = \frac{9}{5}C + 32$ could have been stated as follows:

To convert a Celsius temperature to the equivalent Fahrenheit temperature multiply the Celsius temperature by nine, divide by five and then add thirty-two to your answer.

Clearly the statement $F = \frac{9}{5}C + 32$ is more concise but it does require us to be able to understand this use of letters as a mathematical shorthand.

How did you use the formula to work out the Fahrenheit equivalent of 200°C?

- Perhaps you substituted $C = 200$ into the formula:

$$F = \frac{9}{5}C + 32$$

If $C = 200$,
$$F = \frac{9}{5}(200) + 32$$
$$= 360 + 32$$
$$= 392$$

Note that in this method we need to remember the 'rule of order'. The 200 needs to be multiplied by 9 and divided by 5 **before** we add 32.

- Perhaps you determined the values of F for the various values of C using the ability of some calculators to determine unknown values in a formula given sufficient information.

> Equation:
> $F = \frac{9}{5} \cdot C + 32$
>
> ⦿ F = 392
> ○ C = 200
> Lower = –9E+999
> Upper = 9E+999

EXAMPLE 1

Formula: $A = P + I$

Find A given that $P = 4000$ and $I = 525$.

Solution

Start with the given formula:	$A = P + I$
Substitute in the known values:	$A = 4000 + 525$
Evaluate:	$= 4525$

Thus $A = 4525$.

Note: The formula given in the above example is very straightforward so the above procedure would probably be carried out mentally. However, as before, the same result can be obtained using the ability of some calculators to determine unknown values in a formula given sufficient information.

```
Equation:
A = P + I

⦿ A = 4525
○ P = 4000
○ I = 525
Lower = –9E+999
Upper = 9E+999
```

EXAMPLE 2

Formula: $s = ut + \dfrac{1}{2}at^2$

Find s given that $u = 5$, $t = 3$ and $a = 4$.

Solution

Start with the given formula:

$$s = ut + \frac{1}{2}at^2$$

Substitute in the known values:

$$s = (5)(3) + \frac{1}{2}(4)(3)^2$$

Evaluate:

$$= 15 + \frac{1}{2}(4)(9)$$

$$= 33$$

Thus, when $u = 5$, $t = 3$ and $a = 4$, $s = 33$.

Alternatively use the appropriate facility of some calculators.

EXAMPLE 3

Formula: $A = 250 (1.08)^n$

Find A correct to one decimal place given that $n = 14$.

Solution

Either by using a 'solve facility':

```
Equation:
A = 250·1.08ⁿ
⊙ A = 734.298406064415
○ n = 14
Lower = –9E+999
Upper = 9E+999
```

or 'by hand':

$$A = 250 (1.08)^n$$

$\therefore \qquad A = 250 (1.08)^{14}$

≈ 734.298

Thus $A = 734.3$ correct to one decimal place.

When $n = 14$, $A = 734.3$ correct to 1 decimal place.

In the next example only the 'by hand' method is shown. If you wish to use the 'solve facility' of your calculator instead do still read through the example to check that you understand the process of substituting values into the formula and then check you can obtain the same answer using your calculator.

EXAMPLE 4

A particular industrial process involves the mining and production of a particular metal to various levels of purity. The production of 1 tonne of the metal of $p\%$ purity costs the company $\$C$ where a good approximation of C is given by the formula:

$$C = 1500 + \frac{125\,000}{100 - p}$$

Find the cost of producing one tonne of:

a   the metal of 90% purity,   b   the metal of 99% purity.

Solution

a $C = 1500 + \dfrac{125\,000}{100 - p}$

We are given that $p = 90$, thus $C = 1500 + \dfrac{125\,000}{100 - 90}$

$= 1500 + \dfrac{125\,000}{10}$

$= 14\,000$

To produce one tonne of the metal of 90% purity costs approximately $14 000.

b
$$C = 1500 + \frac{125\,000}{100 - p}$$

We are given that $p = 99$, thus
$$C = 1500 + \frac{125\,000}{100 - 99}$$

$$= 1500 + \frac{125\,000}{1}$$

$$= 126\,500$$

To produce one tonne of the metal of 99% purity costs approximately \$126 500.

Exercise 1A

1 $v = u + at$

   a Find v given that $u = 3$, $a = 5$ and $t = 4$.

   b Find v given that $u = 32$, $a = -2$ and $t = 5$.

   c Find v given that $u = 2$, $a = 20$ and $t = 2.5$.

2 $C = 2\pi r$

   a Find C given that $r = 3$.

   b Find C given that $r = 15$.

   c Find C given that $r = 2.8$.

3 $V = \frac{4}{3}\pi r^3$

   a Find V given that $r = 3$.

   b Find V given that $r = 6$.

   c Find V given that $r = 10$.

4 $s = \frac{u + v}{2} t$

   a Find s given that $u = 3$, $v = 5$ and $t = 10$.

   b Find s given that $u = 12$, $v = -3$ and $t = 8$.

   c Find s given that $u = 22$, $v = 16$ and $t = 7$.

5 $A = 400\,(1.12)^n$

   a Find A given that $n = 5$.

   b Find A given that $n = 8$.

   c Find A given that $n = 10$.

6 A particular industrial process involves the mining and production of a particular metal to various levels of purity. The production of 1 kg of the metal of $p\%$ purity costs the company $\$C$ where a good approximation of C is given by the formula:

$$C = \frac{1000}{100 - p}$$

Find

  a the value of C when $p = 90$,

  b the cost of producing one kg of the metal of 99% purity,

  c the cost of producing one kg of the metal of 99.9% purity.

7 A particular industrial process involves the mining and production of a particular metal to various levels of purity. The production of 1 tonne of the metal of $p\%$ purity costs the company $\$C$ where a good approximation of C is given by the formula:

$$C = 7800 + \frac{80\,000}{100 - p}$$

Find

  a the value of C when $p = 60$,

  b the cost of producing one tonne of the metal of 75% purity,

  c the cost of producing one tonne of the metal of 95% purity.

8 As divers dive to greater depths the pressure on the divers increases due to the weight of the increasing amount of water above them. The pressure, P, in Newtons per square metre, N/m^2, is given by the formula:

$$P = 9.8xd$$

where x metres is how far the diver is below the surface of the water, and d kg/m^3 is the density of the water.

Find, in N/m^2,

  a the pressure at a depth of 5 metres in liquid of density 1000 kg/m^3,

  b the pressure at a depth of 10 metres in liquid of density 1030 kg/m^3,

  c the pressure at a depth of 30 metres in liquid of density 1030 kg/m^3.

9 A scientist suggests that for a particular drug the dose (d units) that should be given to a child, aged c years, can be calculated from the dose (D units) that would be given to an adult, using the rule:

$$d = \frac{(c+1)}{19} D.$$

  a If the adult dose is 15 units, what would be the dose, to the nearest half unit, for a child aged

  i 5 years? **ii** 10 years? **iii** 15 years?

  b The rule makes sense for c up to a certain value. What value is that and why does the rule make no sense after that?

10 To find the density of a solid or liquid we divide its mass by its volume.

$$\text{i.e. Density } (d) = \frac{\text{Mass } (m)}{\text{Volume } (V)}.$$

If the mass is in kilograms and the volume in cubic metres the density will be in kg/m³. If the mass is in grams and the volume in cubic centimetres the density will be in g/cm³, etc.

a Find the density of aluminium if 100 cm³ has a mass of 270 g.

b Find the density of lead if 0.2 m³ has a mass of 2270 kg.

c Find the density of diamond if 0.2 cm³ has a mass of 0.702 g.

The density of sea water increases with depth due to the pressure of the water above.
Find the density of sea water in g/cm³

d just below the surface where 5 cm³ has a mass of 5.141 g,

e at a depth of 1000 m where 5 cm³ has a mass of 5.164 g,

f at a depth of 10 000 m where 5 cm³ has a mass of 5.355 g.

11 If you deposit $P in a bank account that earns interest at x% per annum, compounded annually, the amount in the account after t years will be $A where

$$A = P\left(1+\frac{x}{100}\right)^{t}.$$

Find the amount in the account after 5 years if:

a $300 is deposited at 10% per annum compounded annually,

b $500 is deposited at 12% per annum compounded annually,

c $12 000 is deposited at 7% per annum compounded annually.

12 The pendulum of the clock shown on the right performs one cycle as it swings from A to B and back to A again. The time taken for a pendulum of length l metres to perform one cycle is T seconds where

$$T = 2\pi\sqrt{\frac{l}{9.8}}.$$

Find the value of T (correct to two decimal places) for a pendulum of length

a 1.15 m,

b 40 cm,

c 20 cm.

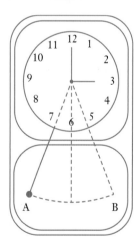

Use of formulae in spreadsheets

In a spreadsheet, entries are made into a table arrangement involving rows and columns. The entries can then be manipulated and used in calculations.

The spreadsheet shown below has values for u, a and t in columns A, B and C respectively, as indicated by the use of row 1 for appropriate labels. The entries in column D are calculated from the values for u, a and t according to the formula $v = u + at$. For cell D2 the formula entered in the cell is: = A2+B2*C2

Note • Spreadsheet formulae always start with =.

• Appropriate formulae for cells D3 to D9 can be entered using the formula in D2 together with the ability of spreadsheets to 'fill down'.

This will place the formula = A3+B3*C3 in cell D3,
= A4+B4*C4 in cell D4, etc.

• Spreadsheets recognise the symbol * as representing multiplication.

	A	B	C	D
1	u	a	t	v
2	5	4	0	5
3	5	4	2	13
4	5	4	4	21
5	5	4	6	29
6	5	4	8	37
7	5	4	10	45
8	5	4	12	53
9	5	4	14	61

= C2+2
= A2+B2*C2
= A3+B3*C3

Create such a spreadsheet yourself for the values of u, a and t shown above.

Spreadsheets can be especially useful for accounting purposes. The spreadsheet below shows the various deposits and withdrawals from a bank account over a period of time. Column E, when complete, will show the balance after each transaction.

	A	B	C	D	E
	Date	Details	Withdrawals ($)	Deposits ($)	Balance ($)
1					
2	1-Mar	Opening balance			1245.65
3	2-Mar	Pay cheque		1243.65	2489.3
4	3-Mar	Rent	480		
5	5-Mar	Shoes	80.9		
6	7-Mar	Supermarket	92.45		
7	8-Mar	Cash out	200		
8	10-Mar	Cheque refund		17.5	
9	10-Mar	Credit card payment	245		
10	14-Mar	Supermarket	185.6		
11	15-Mar	Cash out	200		
12	16-Mar	Pay cheque		1243.65	

Create the above spreadsheet yourself and complete it.

The spreadsheet below shows the marks obtained by 12 students in the ten assessment items of a college course.

	A	B	C	D	E	F	G	H	I	J	K	L
1	Name	Item	Item	Item	Item	Item	Item	Item	Item	Item	Item	Total
2		1	2	3	4	5	6	7	8	9	10	out of 100
3	Sally A	6	7	8	18	17	24	16	11	32	7	
4	Chris C	4	6	6	14	11	14	9	7	26	8	
5	Charlie C	7	10	9	21	19	18	16	17	32	10	
6	Ying H	9	9	7	24	17	23	14	14	37	11	
7	Su J	5	6	6	13	18	17	13	12	23	7	
8	Sanji L	8	9	6	13	16	19	16	17	22	9	
9	Connie N	8	5	6	13	13	14	13	11	22	8	
10	Michele R	7	7	7	18	13	19	10	11	29	9	
11	Becky S	3	5	5	10	12	10	12	9	23	5	
12	Chris T	9	7	10	24	19	22	17	18	33	14	
13	Duane W	10	10	8	20	20	23	17	14	31	15	
14	Icolyn Y	9	8	8	17	14	19	15	16	24	11	
15												
16		10	10	10	25	25	25	20	20	40	16	100

Column L is to show the total mark for each student out of 100, with each of the ten assessment items counting for 10 marks in this total of 100.

Write a suitable formula for cell L3.

Create the above spreadsheet yourself and complete it.

The spreadsheet below is for three items, 1, 2 and 3, that a company sells for $25 each, $30 each and $35 each respectively. The spreadsheet shows orders numbered 6001 (for 3 item 1s, 2 item 2s and 1 item 3), 6002 (for 2 item 1s and 2 item 3s), etc. Notice that column E gives the total cost of the order, column F gives the 10% Goods and Services Tax (GST), column G gives the cost + GST total and columns H, I and J give the remaining stock numbers of each item, starting from the initial stock level of 500 for each item.

	A	B	C	D	E	F	G	H	I	J
1	Order#	Item 1	Item 2	Item 3	Cost	GST	Total	500	500	500
2	6001	3	2	1	$170.00	$17.00	$187.00	497	498	499
3	6002	2	0	2	$120.00	$12.00	$132.00	495	498	497
4	6003	0	1	5	$205.00	$20.50	$225.50	495	497	492
5	6004	2	2	2	$180.00	$18.00	$198.00	493	495	490

Create such a spreadsheet yourself for orders 6001 to 6010 where 6005 to 6010 are as given below.

Order	Item 1	Item 2	Item 3		Order	Item 1	Item 2	Item 3
6005	5	4	5		6006	3	7	4
6007	2	1	6		6008	0	20	11
6009	1	3	3		6010	7	0	9

Miscellaneous exercise one

This exercise may include questions involving the work of this chapter and the ideas mentioned in the Preliminary work at the beginning of the book.

1 Evaluate each of the following:

   a $(-4)^2$ **b** $2 + (-3) \times (-4)$ **c** $2^3 + 3^2$ **d** $(2^3 + 2)^2$

   e $6 - 5 \times (-4)$ **f** $(-1)^5$ **g** $(-1)^6$ **h** $\dfrac{8+4}{4}$

   i $\dfrac{14.14}{10}$ **j** $\dfrac{16.16}{8}$ **k** $1 + (15 - 3 \times 4)^2$ **l** $6 - -4 \div 2$

2 Find the value of each of the following expressions given that $x = 5$.

   a $x + 2$ **b** $2x$ **c** x^2 **d** $2x + 3$

   e $3x + 2$ **f** x^3 **g** $4x - 1$ **h** $1 - 4x$

   i $2(3 + x)$ **j** $(3 + x)^2$ **k** $(x - 3)^3$ **l** $4(2x - 3)$

   m $\dfrac{x+3}{2}$ **n** $\dfrac{4x-2}{3}$ **o** $\dfrac{15}{x-2}$ **p** $\sqrt{3x+1}$

   q $\sqrt{8x-4}$ **r** $\sqrt[3]{2x-2}$

3 Without using your calculator, use appropriate rounding to determine estimates for each of the following, showing your method.

   a 208×84 **b** 19.6×4.7 **c** $\dfrac{208}{9.7}$ **d** $\dfrac{4864}{103}$

   e Attempting to estimate how far it was from one location to another a person counted the number of paces they took in walking from one to the other. They found it required 623 paces. The person measured a typical pace as being approximately 80 cm. Estimate the distance between the two locations.

4 One way of categorising a person as overweight, correct weight or underweight is to consider their 'mass on height squared index', I, also called the body mass index or BMI.

This index is given by $I = \dfrac{m}{h^2}$ where m kg is the mass of the person and h metres is their height.

If $I < 20$, the person is considered underweight, if $20 \le I \le 25$, the person is considered to be the correct weight, and if $I > 25$, the person is considered overweight.

Classify each of the following people

   a Julie, 170 cm, 65 kg,   **b** Alex, 1.95 m, 60 kg,

   c Bill, 195 cm, 97 kg,   **d** Betty, 1.65 m, 55 kg.

Construct a spreadsheet that will automatically calculate the body mass index of a person given their mass and height. Include the four people mentioned above in your spreadsheet.

Challenge: Can you get your spreadsheet to allocate the classifications too?

2.

Percentages

Situation One

The workforce at a particular factory were all given the following letter:

XYZ Company
3A The Industrial Complex, Hope Lane.

To all employees

It is with regret that as managing director I have to inform you that due to the recent political upheaval in a particular country our company no longer has the major supply contracts with this country that we have enjoyed for the past five years. These contracts have fuelled the expansion our company has experienced in recent years.

These lost contracts will have a significant impact on our company and as a consequence the factory will have to operate on a four day week, instead of its usual five day week. Hence, from the first Monday of next month, the factory will not operate on Fridays. For the period of time for which this four day week is necessary all employees will receive a reduced salary equal to their old salary less 20%.

We regret this course of action but feel it necessary for the company to survive financially with the reduced order book. Rest assured that every effort will be made to find replacement orders and when that occurs the factory will return to a five day week.

Yours sincerely

Managing director.

Some weeks later they received a second letter:

XYZ Company
3A The Industrial Complex, Hope Lane.

To all employees

Great news — Due to the hard work of our sales team we have won several new contracts recently and, in addition, the new government in the country that recently cancelled its long term orders with us has restored them. Hence we can return to our five day week as from next Monday.

The salary we have all been existing on in recent weeks will now be increased by 20% to take it back to what it was before the forced reduction.

I wish to pass on my thanks to you all for the understanding you have shown in the recent difficulties.

Yours sincerely

Managing director.

Write a note to the managing director regarding this second note.

Mr Chan is very concerned to read in a newspaper article that a particular drug that his wife needs to take for a medical condition she suffers from has:

'… the potential to increase the likelihood of a person taking the drug from developing a particular form of cancer by up to 50% of the risk for someone not taking the drug.'

The article also said that

'… for women not taking this drug there was a 1% chance of developing this cancer.'

Mr Chan was very concerned to read that this latest research seemed to indicate that the risk to his wife could in fact be as high as 51%!

Is there anything that could be said to calm Mr Chan's worries?

We will now revise

- expressing an amount as a percentage of some total amount,

and • finding a percentage of something.

It is anticipated you will have encountered both of these ideas before, as mentioned in the *Preliminary work* section at the beginning of this book.

Expressing an amount as a percentage of some total amount

EXAMPLE 1

Express each of the following as a percentage.

a   18 students out of a total of 40 students.

b $3.60 out of $45.00.

c $34 out of a total of $245.

Solution

a   Express 18 out of 40 as a fraction: $\dfrac{18}{40}$

Find this fraction of 100: $\dfrac{18}{40} \times 100 = 45$

18 students out of 40 is 45%.

```
18 / 40 × 100
                    45
```

b Express $3.60 out of $45 as a fraction: $\dfrac{3.60}{45}$

Find this fraction of 100: $\dfrac{3.60}{45} \times 100 = 8$

$3.60 out of $45.00 is 8%.

> 3.6 / 45 × 100
>   8

c Express $34 out of $245 as a fraction: $\dfrac{34}{245}$

Find this fraction of 100: $\dfrac{34}{245} \times 100 \approx 13.9$

$34 out of a total of $245 is 13.9% (correct to 1 decimal place).

> 34 / 245 × 100
>   13.87755102

Expressing an increase or decrease as a percentage of something

If asked to express an increase (or decrease) as a percentage increase (or decrease), give the increase (or decrease) as a percentage of the *original* amount.

The price of a commodity increases from $1640 per tonne to $1846 per tonne. Express the increase as a percentage increase.

Solution

$$\text{Increase} = \$1846 - \$1640$$
$$= \$206$$

As a fraction of the original amount: $\dfrac{\$206}{\$1640}$

Find this fraction of 100: $\dfrac{\$206}{\$1640} \times 100 \approx 12.56$

> 206 / 1640 × 100
>   12.56097561

The price of the commodity has increased by 12.6%, correct to 1 decimal place.

Finding a percentage of something

It is anticipated that you have previously encountered:

- finding percentages of amounts (see **example 3** that follows),

- increasing or decreasing something by a given percentage (see **example 4**),

- working back to find an original quantity knowing how much it has become after a given increase or decrease (see **example 5**).

Percentage shortcuts

Working with percentages

EXAMPLE 3

Find 27% of $60.

Solution

Method 1
(Find 1% and then find 27%.)

$$1\% \text{ of } \$60 = \frac{\$60}{100}$$

$$\therefore 27\% \text{ of } \$60 = \frac{\$60}{100} \times 27$$
$$= \$16.20$$

Method 2
(Use the decimal equivalent of 27%.)

$$27\% \text{ of } \$60 = \$60 \times 0.27$$
$$= \$16.20$$

EXAMPLE 4

Increase $450 by 24%.

Solution

Method 1
(Find 24%, then increase.)

$$1\% \text{ of } \$450 = \frac{\$450}{100}$$

$$\therefore 24\% \text{ of } \$450 = \frac{\$450}{100} \times 24$$
$$= \$108$$

$$\$450 + \$108 = \$558$$

$450 increased by 24% is $558.

Method 2
(Use the decimal equivalent of 124%)

$$124\% \text{ of } \$450 = \$450 \times 1.24$$
$$= \$558$$

EXAMPLE 5

After the 10% goods and services tax (GST) has been added to an invoice, the final amount payable is $522.50. Determine the amount on the invoice prior to the GST being added.

Solution

Method 1 (Find 1% and then find 100%.)
After 10% has been added we have 110% of the pre GST invoice.

Thus   110% of pre GST amount = $522.50

$$\therefore \qquad 1\% \text{ of pre GST amount} = \frac{\$522.50}{110}$$

$$\text{Hence} \qquad 100\% \text{ of pre GST amount} = \frac{\$522.50}{110} \times 100$$
$$= \$475$$

The amount on the invoice prior to the GST being added was $475.

Method 2 (Use the decimal equivalent of a 10% increase.)
An increase of 10% gives us '1.1 times' the original amount.

Thus $1.1 \times$ pre GST amount = $522.50

$$\therefore \qquad \text{Amount prior to GST} = \frac{\$522.50}{1.1}$$
$$= \$475$$

As before: The amount on the invoice prior to the GST being added was $475.

ISBN 9780170390194

Exercise 2A

1 What number would you multiply an amount by so that your answer
  - **a** is 10% of the amount
  - **b** is 30% of the amount
  - **c** is 25% of the amount
  - **d** is 4% of the amount
  - **e** is 12.5% of the amount
  - **f** has increased the amount by 40%
  - **g** has increased the amount by 10%
  - **h** has increased the amount by 23%
  - **i** has increased the amount by 4%
  - **j** has increased the amount by 12.5%
  - **k** has decreased the amount by 10%
  - **l** has decreased the amount by 8%
  - **m** has decreased the amount by 18%
  - **n** has decreased the amount by 60%
  - **o** has decreased the amount by 2.5%?

2 Express each of the following as a percentage.
  (Give your answer to the nearest 0.1% if rounding is necessary.)
  - **a** 21 students out of a total of 50 students.
  - **b** $18 out of $25.
  - **c** $2.25 out of a total of $18.
  - **d** 174 sheep out of a total of 1356 sheep.
  - **e** 8.5 cm out of a total of 2.5 metres.
  - **f** 35 metres out of a total of 5.832 km.

3 Find
  - **a** 20% of $100
  - **b** 30% of $200
  - **c** 10% of $60
  - **d** 25% of $88
  - **e** 16% of 250 kg
  - **f** 55% of $23
  - **g** 12% of 5 metres
  - **h** 45% of 3 tonnes

4 Increase
  - **a** $40 by 50%
  - **b** $80 by 20%
  - **c** 160 kg by 10%
  - **d** 55 metres by 40%
  - **e** $23.45 by 80%
  - **f** $500 by 17%
  - **g** 60 litres by 5%
  - **h** $250 by 12.5%

Percentages without
calculators

5 Decrease
  - **a** $80 by 50%
  - **b** $18 by 25%
  - **c** 540 kg by 25%
  - **d** 6 metres by 2%
  - **e** 6.5 metres by 20%
  - **f** $23 by 90%
  - **g** $88 by 12.5%
  - **h** 160 tonnes by 45%

Repeated percentage
change

6 a Find 23% of $124.60 giving your answer to

    i the nearest cent   **ii** the nearest five cents   **iii** the nearest ten cents.

b Increase $1260 by 14.5% giving your answer to the nearest dollar.

c Decrease $1260 by 14.5% giving your answer to the nearest dollar.

7 a 1% of an amount is $13.45, find the amount.

b 15% of an amount is $60, find the amount.

c 45% of an amount is $117, find the amount.

d 28% of an amount is $44.38, find the amount.

e 11.8% of an amount is $1479.72, find the amount.

f After a 10% price rise an item has a price of $268.40. What was the price of the item before the rise?

g After a 4% rise in the value of some shares the shares were worth $1348.15. To the nearest dollar what were they worth before the 4% rise?

h In a sale all normal prices are reduced by 15%. Find the normal price of an item with a sale price of $39.95.

i In a sale all normal prices are reduced by 12.5%. Find the normal price of an item with a sale price of $112.00.

j After a pay rise of 5% Joe's weekly pay rises to $876.75. What was Joe's weekly pay before the rise?

k A 10% monthly drop in sales saw the number of new cars sold by a particular manufacturer fall to 2106 in one month. How many cars did this manufacturer sell in the previous month?

8 The price of a commodity increases from $796 per m^3 to $873 per m^3. Express the increase as a percentage increase.

9 The share price for a particular company falls from $72.54 to $67.92. Express this decrease as a percentage decrease.

10 Approximately 46% of a class of 26 students are boys. How many girls are in the class?

11 The 14 girls in a class of students form approximately 44% of the students in the class. How many students are there in the class?

12 In earlier years, legal documents required an official government 'stamp' to be attached to them, or impressed upon then, to make the document legal. Modern documents often do not require this but the government may still charge a 'stamp duty' to register the document. Suppose the stamp duty payable on the transfer of a property sold for $650 000 is charged at $11 200 + 4.7% of the amount the sale price exceeds $360 000. How much stamp duty is payable?

13 Copy and complete the following table:

	Number of items	Cost per item	Sub total	GST (10%)	Total
e.g.	15	$16.40	$246.00	$24.60	$270.60
a	23	$17.50			
b	131	$16.40			
c	18	$15.90			
d	24		$420.00		
e		$19.85	$119.10		
f	15		$1129.50		
g	26			$20.80	
h		$6.75		$9.45	
i		$3.40			$463.76
j	18				$767.25

14 A factory increased its annual output of new cars from 1219 in one year to 1317 in the next. Express this increase as a percentage increase.

15 A person is given a 4.8% pay rise. If the person was earning $1971.15 per fortnight before the rise what will they earn per fortnight after the rise?

16 The rainfall total for a particular region for the year 2007 was 11% down on the total for 2006. The 2006 total was 254 mm. What was the rainfall total for this region in 2007, to the nearest mm?

17 a The child dose of a particular medicine is 50% of the adult dose. If the adult dose is 20 milligrams what is the child dose?

b The child dose of a particular medicine is 40% of the adult dose. If the child dose is 5 millilitres what is the adult dose?

18 Toni bought a house for $315 200 and sold it some years later for $475 600. What was the percentage increase in the cost of the house in this time?

19 A person is earning $1432.23 per fortnight. What percentage increase in earnings is required to take this fortnightly pay to at least $1500.00? (Give your answer to 1 decimal place.)

20 A hardware shop advertises '15% off all marked prices'. With this reduction what should you expect to pay for each of the following with marked prices as indicated?

Drill

Marked price
$125.00

Chain saw

Marked price
$327.00

Sander

Marked price
$36.00

Tool box

Marked price
$28.00

21 The table below shows one scheme for determining how the income tax payable could be worked out from a person's taxable income.

Taxable income	Income tax to pay
$0 → $19400	Nil
$19401 → $37000	19% of the taxable income over $19400
$37001 → $80000	$3344 + 33% of the taxable income over $37000
$80001 → $180000	$17534 + 37% of the taxable income over $80000
$180001 and over	$54534 + 45% of the taxable income over $180000

a According to this table how much income tax will each of the following people be required to pay?

Aimee, taxable income	$17900	Brittney, taxable income	$34400
Chris, taxable income	$38700	Devi, taxable income	$63000
Emily, taxable income	$97200	Frank, taxable income	$213400

b Megan pays $27006 income tax. Calculate Megan's taxable income.

c Allen pays $9878 income tax. Calculate Allen's taxable income.

22 A particular region of Australia produced approximately 146000 kg of apples in one year. This total was 10% up on the number of kilograms produced the previous year which was in turn 8% up on the number produced the year before that.

How many kilograms of apples did the region produce in each of these two previous years?

23 A certain tax rate was increased from 10% of the pre-tax amount to 12.5% of the pre-tax amount. For one item this increased the tax on the item by 440 cents. Determine the pretax amount for this item.

ISBN 9780170390194

Inflation

Inflation is the rise in the price of goods over a period of time. Inflation is usually quoted as an annual percentage rate based on the increase in the price of a 'package' of goods and services as typically used by a household. If, for example, this package were to increase in price during a twelve month period by 3.1% then the annual inflation rate would be quoted as 3.1%.

If the rate at which a person's wage increases is below the inflation rate then the wage is not 'keeping up with inflation'.

In an inflationary environment each dollar earned has less 'purchasing power' than a dollar earned a year earlier. For example, because of the inflation over the last fifty years the quantities of things like meat and vegetables that could be bought for $1 fifty years ago could not be bought for just $1 now. The $1 has lost purchasing power.

> **RESEARCH**
>
> Using internet resources as appropriate find recent inflation rates for Australia and four other countries of your choosing.

Exercise 2B

1 If we assume that the annual rate of inflation were to remain steady at 3.4%, and a particular item costs $45 now, what would this suggest we would be paying for the same item in

   a   one year?   **b**   two years?   **c**   three years?

If the current rate of inflation was indeed 3.4%, how accurate do you feel the predictions you have just calculated are likely to be?

2 Over a five-year period, a country experiences annual inflation rates as follows:

   Year 1: 3.2%   Year 2: 4.3%   Year 3: 5.1%   Year 4: 4.1%   Year 5: 3.3%

Noticing that these rates add up to 20% Jamie concludes that goods that cost $100 at the start of this five-year period would cost $120 at the end of the five years. Is Jamie correct in his conclusion?

3 We would generally expect that something costing $100 in the year 2000 would have cost more by the year 2010 (unless changes in availability and manufacturing had made producing the item cheaper). The expected 2010 cost of something costing $100 in 2000 can be calculated using what is called the *Consumer Price Index* (CPI for short) for each year and by evaluating:

$$\$100 \times \frac{\text{CPI for the year 2010}}{\text{CPI for the year 2000}}.$$

Given the following CPI figures:

   $\text{CPI}_{1950} = 7.85$ $\text{CPI}_{1990} = 103.175$ $\text{CPI}_{2000} = 128.4$ $\text{CPI}_{2010} = 172.6$

According to these figures what, to the nearest dollar, was the cost in 2010, of something costing

   a $400 in 1950?   **b** $50 in 1990?   **c** $1700 in 2000?

> **RESEARCH**
>
> To compare a worker's purchasing power in different countries one suggestion is to use the amount of time it takes a worker, earning that country's average wage, to earn enough to purchase a Big Mac at their local McDonald's store. Research this *Working time Big Mac Index* on the internet.

Goods and Services Tax

When the goods and services tax (GST) was first introduced into Australia business owners and company accounts personnel had to familiarise themselves with the various forms and procedures that accompanied its introduction.

Goods and services tax is added at the rate of 10%. Businesses were told that if they knew the total amount received from selling goods on which 10% GST had been added they had to divide this total by 11 to determine the amount of GST included. This amount had then to be forwarded to the government tax office after any allowable deductions had been made.

Government advisers who were assisting businesses to become familiar with the various requirements associated with the new system reported that a number of people were querying the division by 11, feeling sure it was an error and that as the GST rate was 10% (i.e. one tenth, not one eleventh) they should be dividing by 10 to determine the GST amount in their total takings and not dividing by 11.

Write a letter that could be sent out in response to any such queries explaining why the division by 11 is correct.

Discount

A discount is a reduction of the usual price for some reason and is usually stated as a percentage of the usual price. This discount might be due to a sale, as was the case in some of the questions of an earlier exercise, or due to customer loyalty, bulk purchases, damaged goods, cash purchases, etc.

EXAMPLE 6

A shop offers 5% discount on all goods purchased with cash rather than credit cards.

a   What is the cash discount price for a jacket usually costing $85.00?

b   If a computer game has a discounted price of $159.60 for cash what is the usual price of the computer game without the discount for cash?

Solution

a   With a 5% discount for cash the cash price will be 95% of the usual price.

Hence: $\qquad$ Cash price $= 0.95 \times \$85.00$

$\qquad\qquad\qquad\qquad = \$80.75$

With the discount for cash the price of the jacket is $80.75.

b   With a 5% discount for cash the cash price will be 95% of the usual price.

Hence $\qquad 0.95 \times$ Usual price $= \$159.60$

$\therefore \qquad\qquad$ Usual price $= \dfrac{\$159.60}{0.95}$

$\qquad\qquad\qquad\qquad = \$168.00$

The price of the computer game without the discount for cash is $168.00.

ISBN 9780170390194

Commission

The fees charged by some businesses for selling an item is sometimes based on the sale price achieved. Similarly the amount of pay some people receive is sometimes based on the quantity of sales they achieve.

These *commissions* are often a percentage of the value of the sales.

For example, when a real estate agency sells a property they may receive a commission equal to 2.5% of the amount the property sells for. Thus for selling a property for $640 000 the agency receives $0.025 \times \$640\,000$, i.e. $16 000.

For a computer sales person who is paid commission of 15% of the value of the sales they make in a month, monthly sales of $17 650 would produce earnings for that month of $0.15 \times \$17\,650$, i.e. $2647.50

In some cases the commission may be 'stepped'. Consider for example the real estate agency that charges commission on the sale of a house on the following stepped basis:

Commission: 3% of the first $100 000
2.5% of the next $150 000
2% of the next $150 000
1% thereafter.

For selling a house for $365 000 the agency receives:

$0.03 \times \$100\,000\ +$
$0.025 \times \$150\,000\ +$
$0.02 \times \$115\,000$
for a total of $9050.

For a salesperson whose wage consists entirely of commission earned, a poor month of sales can result in a low wage that month and possible problems trying to meet their other financial commitments such as the food bill, rent payments, phone bill etc. To avoid this situation a person may be paid a fixed wage, or *retainer*, and then a commission 'on top'.

For example, suppose that Jin, a salesperson of new cars, is paid a monthly retainer, equal to $2780, plus commission of 2% of monthly sales.

In a month when Jin's monthly sales total $181 200 Jin receives:

$$\$2780 + 0.02 \times \$181\,200 = \$6404.$$

Profit and loss

If we sell something for more than we paid for it we make a **profit** and if we sell something for less than we paid for it we make a **loss**.

If we buy something for $40 and sell it for $70 our profit is $30. Similarly, if we buy something for $300 and sell it for $330 we make a profit of $30. However, in the first case we made our profit on an initial outlay of just $40 compared to an initial outlay of $300 in the second case. To compare the two we could express each as a **percentage profit**, a term explained on the next page.

ws
Profit and loss

$$\text{Percentage profit} = \frac{\text{Profit made when you sell the item}}{\text{Amount you paid for the item}} \times 100$$

and similarly:

$$\text{Percentage loss} = \frac{\text{Loss made when you sell the item}}{\text{Amount you paid for the item}} \times 100$$

Thus: Buying for \$40 and selling for \$70:

$$\text{Percentage profit} = \frac{30}{40} \times 100$$
$$= 75\%$$

Buying for \$300 and selling for \$330:

$$\text{Percentage profit} = \frac{30}{300} \times 100$$
$$= 10\%$$

Buying for \$250 and selling for \$180:

$$\text{Percentage loss} = \frac{70}{250} \times 100$$
$$= 28\%$$

Exercise 2C
Discount

1 An office furniture shop offers 8% discount on all cash sales. Find the discounted price on each of the following, rounding to the nearest 5 cents if rounding is necessary.

Usual price	Usual price	Usual price	Usual price
$50.00	$67.30	$72.40	$86.90

2 After a discount of 15% an item is priced at \$108.80. Determine the price of the item before the discount.

3 What percentage discount is needed to see a normal price of \$75 reduced to \$67.50?

4 A company offers an 8% discount on all items ordered online from its website ordering facility. What will be the discounted price of an item usually costing \$48.50?

5 A company offers a discount of 6% off the total amount, on all orders over \$500. Under this scheme what would be the price of each of the following orders?

Order One	Order Two	Order Three
26 items at $72.80 each	1 item at $72.80 each	3 items at $72.80 each
58 items at $67.40 each	2 items at $67.40 each	5 items at $67.40 each
137 items at $17.50 each	13 items at $17.50 each	21 items at $17.50 each

6 A manufacturer normally sells a particular item for $56 each. However, to encourage shopkeepers to buy the item and sell them in their shops the manufacturer offers any shopkeeper buying 200 of these items a 20% discount for 'bulk', i.e. a discount for the large number purchased. If a shopkeeper buys 200 at the bulk discounted price and then sells each item at the normal price of $56 how much profit will the manufacturer make

   a   on each item?

   b   on the sale of the entire 200?

   c   The shopkeeper is concerned that he may not sell all of the 200 items. How many does the shopkeeper need to sell to cover what the 200 cost him?

Commission

7 A real estate agency charges commission of 2.5% of the sale price of any property it sells. What commission does this agency charge for the sale of a property for $520 000?

8 A salesperson is paid purely on commission, earning 18% of the total value of the goods he sells. How much does this person earn for a month in which his total sales are $17 800?

9 A salesperson is paid a monthly retainer of $3200 plus a commission of 4% of the total value of all goods sold in that month. How much do they earn in a month when sales total $24 500?

10 A real estate agent is paid 0.8% commission on all sales. Determine the total amount the agent is paid for selling three properties:

          One for $320 000, one for $480 000 and one for $540 000.

11 A salesperson is paid $480 per fortnight plus 8% of the amount by which their total sales for the fortnight exceed $5000. In one fortnight the salesperson sold three items for $4280 each, one item for $960 and four items for $3470 each.

   What was the salesperson's total pay for this fortnight?

12 A person is paid a retainer of $1230 per fortnight plus commission of 3.5% of all goods sold the previous fortnight. One fortnight the person was paid a total of $1868.40. What was the total value of the sales the previous fortnight?

13 A financial adviser charges her clients for managing their share portfolios. The charge is on a commission basis dependent upon the total value of the portfolio and according to the following structure.

For any portfolio with a total value under $150 000:   Fixed fee of $3000.

For portfolios worth a total value of $150 000 or more: 2% of first $250 000
                                               1.5% of the next $250 000
                                             1.2% of the next $100 000
                                                1% thereafter.

Determine the commission charged on portfolios with a total value of

   a $140 000   **b** $180 000   **c** $475 000   **d** $1 567 000

Profit and loss

For numbers 14 to 25 copy and complete the table.

	What it cost	What it was sold for	Profit as percentage of cost
14	$100	$124	
15	$400	$418	
16	$100		18%
17	$650		30%
18		$135	8%
19		$20 625	65%

	What it cost	What it was sold for	Loss as percentage of cost
20	$100	$84	
21	$175	$105	
22	$6500		6%
23	$18.50		20%
24		$29.25	10%
25		$11 132	8%

26 Which shows the greater percentage profit:

Item A purchased for $56 and sold for $72

or

Item B purchased for $3210 and sold for a profit of $910?

27 Meta purchases an item and sells it on to Susan at a profit of 40%.

When Susan sells it to Chris for $1155 she makes a profit of 10%.

How much did Meta pay for the item?

28 Toni purchases things at auction and then sells them in his antiques shop for a profit. The profit he makes on each item varies but he always attempts to make a profit in the 20% to 40% range. At one auction he purchases three items, one for $85, one for $155 and one for $2150. Given that when he sold each one he did make a profit in his desired range what price might he have sold each item for?

29 Jack purchases an item and sells it on to Steve at a profit of 10%.

Steve sells the item on to Nyuma at a profit of 20%.

Nyuma sells it on to Shan for $3795 giving Nyuma a profit of 15%.

How much did Jack pay for the item?

ISBN 9780170390194

Miscellaneous exercise two

This miscellaneous exercise may include questions involving the work of this chapter, the work of any previous chapters, and the ideas mentioned in the Preliminary work at the beginning of the book.

1 Find the value of each of the following expressions given that $x = 5$ and $y = 7$.

a	$x + y$		**b**	$y + x$		**c**	xy
d	$x + 2y$		**e**	$2x + y$		**f**	$2(x + y)$
g	$x + y^2$		**h**	$x^2 y$		**i**	$(x + y)^2$
j	$x^2 + y^2$		**k**	$(x - y)^2$		**l**	$xy - 3x + y$

2 What number do we multiply an amount by if we wish to

a increase the amount by 20%?   **b** find 20% of the amount?

c decrease the amount by 20%?   **d** find 2% of the amount?

e decrease the amount by 2%?   **f** increase the amount by 2%?

3 List each of the following amounts in order, largest first.

A: 35% of \$1500.   B: \$800 increased by 5%.

C: \$750 decreased by 20%.   D: 80% of \$560.

E: \$360 increased by 95%.   F: \$400 decreased by 10%.

4 If a shopkeeper agrees to buy large quantities of an item from a supplier the shopkeeper may receive a 'bulk discount' i.e. a discount due to the large quantity ordered. Suppose a particular item usually costs \$17.50 from the supplier but by purchasing 250 of these a shopkeeper is given a 20% bulk discount. If the shopkeeper then sells each of these at the manufacturer's usual price of \$17.50 what is the profit the shopkeeper is making on each item as a percentage of the amount each item cost him?

If this question had been asked without the quantities '250' and '\$17.50' being given, i.e. only the 20% bulk discount being known, could the answer still have been determined?

5 A high school newsletter states that 18% of its students are in year 8.

If the school has 1643 students altogether and the 18% is given to the nearest whole percentage how many year 8 students are there in the school?

6 A bookshop is charged \$15.40 for a particular book. The shopkeeper wishes to price the book such that his profit would be 30% of his selling price. What should be his selling price?

7 Will an annual inflation rate of 5% mean that in a ten year period the cost of living will rise by 50%? Explain your answer.

8 The table on the right shows the mother's age and the sex of the baby for 1247 babies born during one year at a maternity hospital.

		Age of mother			
		Under 25	25 to 35	Over 35	Totals
Sex of baby	Male	172	403	61	636
	Female	165	384	62	611
	Totals	337	787	123	1247

a How many of the 1247 babies were males born to mothers aged 25–35?

b How many of the 1247 babies were born to mothers aged 25 and over?

c What percentage of babies born to mothers aged under 25 were girls?

9 a The amount of stamp duty that is payable is based on the 'dutiable value' of an item. Let us suppose that the calculation of the amount payable is based on the following structure:

Dutiable value	Stamp duty payable
$0 to $120 000	1.9% of dutiable value.
$120 001 to $150 000	$2280 + 2.85% of dutiable value over $120 000
$150 001 to $360 000	$3135 + 3.80% of dutiable value over $150 000
$360 001 to $720 000	$11 115 + 4.75% of dutiable value over $360 000
Over $720 000	$28 215 + 5.15% of dutiable value over $720 000

(Note how there are no 'sudden jumps' when moving from one line to another.)

Determine the stamp duty payable based on a dutiable value of

i $6000   **ii** $125 000   **iii** $380 000   **iv** $870 000

b A similar system of stamp duty payments is to be set up according to the structure given below. Copy and complete the structure filling in the blanks with the appropriate figures.

Dutiable value	Stamp duty payable
$0 to $150 000	1.5% of dutiable value.
$150 001 to $300 000	$____ + 2.8% of dutiable value over $150 000
$300 001 to $500 000	$____ + 3.5% of dutiable value over $300 000
$500 001 to $750 000	$____ + ___% of dutiable value over $500 000
Over $750 000	$23 950 + 5.1% of dutiable value over $750 000

10 When a house purchased by Jack is sold to Jill, Jack makes a profit of 80%.

Jill later sells the house to Samantha and this time Jill makes a profit of 140%.
If Samantha paid $777 600 for the house how much did Jill and Jack pay for it?

11 The truck shown sketched is used to transport cement. The container can safely be filled to $0.4V$ where V, the capacity of the container, is given by:

$$V = \frac{\pi}{3}(a^2x + b^2x + abx + 2b^3)$$

with a, b and x in linear units and V in those linear units cubed.

If $a = 1$ metre, $b = 1.2$ metres and $x = 3$ metres find V and state the units. What volume of cement can the container safely hold?

3.

Simple interest

Investing money

Suppose we deposit $500 into a savings account that earns interest at the rate of 10% per year ('per year' is often referred to as *per annum*). How much will this account be worth after 1 year? How much will this account be worth after 2 years? Suppose we take the money out of the account after just six months, or just a few days?

To answer these questions we would need to know how the account operates and, in particular, how the interest is calculated and when it is added to the account. In this chapter we will consider the process called 'simple interest' and in the next chapter we will consider the 'compound interest' process.

Simple interest

Suppose the $500 is invested in an account for which the interest is paid at 10% per annum *simple interest*. This means that each year, 10% of $500 is added to the account as interest.

Simple interest 1

Simple interest 2

Thus after 1 year the account will be worth
$$\begin{aligned} & \$500 + 10\% \text{ of } \$500 \\ = \ & \$500 + \$50 \\ = \ & \$550 \end{aligned}$$

After 2 years the account will be worth
$$\begin{aligned} & \$500 + 2 \times 10\% \text{ of } \$500 \\ = \ & \$500 + 2 \times \$50 \\ = \ & \$600 \end{aligned}$$

After 3 years the account will be worth
$$\begin{aligned} & \$500 + 3 \times 10\% \text{ of } \$500 \\ = \ & \$500 + 3 \times \$50 \\ = \ & \$650 \end{aligned}$$

Each year the account is worth $50 more than the previous year:

Value at end of one year = Value at start of that year + $50

If we want the value after n years:

Value after n years = Initial value + $n \times \$50$

Notice that whilst the interest is earned each year it is not added to the account until some later date (e.g. when the account is closed). In this way the interest from the first year does not itself earn interest in the second year. This is what distinguishes *simple interest*, considered in this chapter, with *compound interest* considered in the next chapter. In compound interest, interest is earned on the interest.

A spreadsheet could be used to determine and display the value of the account after 1, 2, 3, … years, as shown on the next page.

	A	B	C	D	E	F	G	H
1	Amount invested ($)		$500.00					
2	Annual interest rate		10.00%					
3		Simple interest account.						
4	End of year	1	$550.00					
5		2	$600.00					
6		3	$650.00					
7		4	$700.00					
8		5	$750.00					
9		6	$800.00					
10		7	$850.00					
11		8	$900.00					

Try to create such a spreadsheet for which you can simply enter different values for the amount invested, cell C1 in the above spreadsheet, and the percentage rate, cell C2 in the above spreadsheet, and the spreadsheet automatically recalculates the amount for the end of each year. (If you do manage to create such a spreadsheet, do not delete it as we will use it again in the next chapter when considering compound interest.)

EXAMPLE 1

Determine the interest paid on $2500 invested at 8% per annum simple interest for 3 years. How much is the investment worth at the end of the three years?

Solution

Interest each year is 8% of $2500 = $2500 × 0.08
$$= \$200$$

Thus after 3 years the interest is $600.

Thus at the end of the three years the investment will be worth $2500 + $600
$$= \$3100.$$

Alternatively we could use the built-in capability of some calculators, or some computer or internet programs, to perform simple interest calculations.

The simple interest formula

If we were to invest $P at R% per annum simple interest:

Interest received for one year $= \$P \times \dfrac{R}{100}$

After T years, the total interest $= \$P \times \dfrac{R}{100} \times T.$

Thus a *principal* of $P invested for T years at R% simple interest p.a. (per annum) earns interest of $I where

$$I = \frac{PRT}{100}$$

ISBN 9780170390194

If instead we use R in decimal form, not as a percentage, the formula becomes

$$I = PRT$$

For example, if the interest rate were 6% the first formula would use $R = 6$ but the second formula would use $R = 0.06$.

EXAMPLE 2

Determine the interest paid on $8400 invested at 2.4% per annum simple interest for 5 years.

Solution

Using $I = PRT$
Interest $= \$8400 \times 0.024 \times 5$
$= \$1008$
The interest paid is $1008.

```
8400 × 0.024 × 5
                      1008
```

EXAMPLE 3

Determine the value after two years of an investment of $75 000 invested at 4.6% per annum simple interest.

Solution

Using $I = PRT$
Interest $= \$75\,000 \times 0.046 \times 2$
$= \$6900$
$\therefore$ Value after 2 years $= \$75\,000 + \$6900$
$= \$81\,900$
The value after two years is $81 900.

```
75000 × 0.046 × 2
                      6900
```

Daily, monthly, quarterly and six-monthly interest rates

Suppose that T, the time a sum of money is invested for, is given as a number of days, or months, or quarters (a 'quarter' is a quarter of a year, i.e. three months), or half year periods rather than as a number of years. If the annual interest rate is, for example, 6% then when using $I = PRT$:

With T in days we use

$$I = P \times \frac{0.06}{365} \times T$$

With T in months we use

$$I = P \times \frac{0.06}{12} \times T$$

With T in quarters we use

$$I = P \times \frac{0.06}{4} \times T$$

With T in 6-month periods we use

$$I = P \times \frac{0.06}{2} \times T$$

Note • For convenience, in this book, we will ignore leap years and assume that all years have 365 days.

• Some banking calculations, again for convenience, are based on a concept called a *banker's year*. This concept takes a year as being 360 days and consisting of twelve equal months each of thirty days.

EXAMPLE 4

How much interest is due after 56 days for an investment of $24 275 invested at 5% per annum? (Give your answer to the nearest cent.)

Solution

Using $I = PRT$

$$\text{Interest} = \$24\,275 \times \frac{0.05}{365} \times 56$$

$$= \$186.22 \text{ to the nearest cent.}$$

The interest due is $186.22, to the nearest cent.

EXAMPLE 5

How much interest is due after 10 months for an investment of $125 000 invested at 4.7% per annum? (Give your answer to the nearest cent.)

Solution

Using $I = PRT$

$$\text{Interest} = \$125\,000 \times \frac{0.047}{12} \times 10$$

$$= \$4895.83 \text{ to the nearest cent.}$$

The interest due is $4895.83, to the nearest cent.

Exercise 3A

1 Determine the interest paid on $4000 invested at 5% per annum simple interest for 2 years.

2 Determine the interest earned and the final value of the investment if $500 is invested at 12% per annum simple interest for 6 years.

3 How much will an account be worth after 15 years if $5000 is invested at 4.8% per annum simple interest?

4 A savings account paying five percent per annum simple interest is opened with an initial deposit of eight thousand seven hundred dollars. If no further deposits are made, how much will this account be worth when it is closed seven years later?

5 Julie is left a sum of $3450 from her late aunt's will. She decides to spend $1000 of this amount on a holiday and invest the rest in an account paying 6.2% per annum simple interest.

    a   How much will the account be worth three years later?

    b   How much more would the account have been worth had Julie invested the whole $3450 for the three years?

6 Jack invests $5400 for three years in an account paying 4.5% per annum simple interest. How much more interest would he have received had he instead invested the same amount for the same time but in an account paying 5.4% per annum simple interest?

7 How much interest is earned when one million dollars is invested for one month in an account paying 5.4% per annum simple interest?

8 Suppose you were allowed to use one billion dollars ($1 000 000 000) for one day. How much interest would you receive if you invested the billion for the day in an account paying 6.2% per annum simple interest?

9 Shaq invests $750 in an account paying 7.5% per annum simple interest for three quarters of a year. How much will the account be worth at the end of this time, to the nearest cent?

10 How much interest is earned when $3500 is invested for 13 months in an account paying 9.8% per annum simple interest?

11 $52 000 is invested for six months in an account paying 6.95% per annum simple interest. What will be the value of this account at the end of this time?

12 Penny has $8300 to invest. The bank advises her that they have an account that normally pays 5.4% per annum simple interest but at the moment this account is offering an extra 1% per annum, to make the special offer interest rate of 6.4% per annum simple interest. If she invests the money for 200 days how much extra interest will the special offer give her in this time, over the normal rate?

13 How much interest accrues if $17 140 is invested for sixty days in an account paying 11.7% per annum simple interest?

The questions in this exercise involved simple interest rates that ranged from a low of 4.5% per annum to a high of 12% per annum. In reality, if you attempted to open a savings account today, is there likely to be that much variation in the various accounts that the various banks and financial institutions offer? Explain your answer.

Consider the table on the right showing the transactions and the balances for a bank savings account for the month of April and May.

Let us suppose that interest is paid on this account and the interest for April and May will be added on the first day of June. Hence interest for April will *not* itself earn interest in May.

If interest is calculated at the rate of 6% per annum how much interest would the account earn for the month of April and for the month of May?

To be able to answer this question we need to know what method the bank uses to calculate the interest. Here we will consider two methods:

- Minimum monthly balance,

and   • Daily balance.

Date	Details	Amount	Balance
01 April	Opening balance		$250.00
07 April	Deposit	$500.00	$750.00
11 April	Cash withdrawal	$350.00	$400.00
21 April	Deposit	$500.00	$900.00
30 April	Final balance		$900.00

Date	Details	Amount	Balance
01 May	Opening balance		$900.00
05 May	Deposit	$500.00	$1400.00
13 May	Cash withdrawal	$800.00	$600.00
19 May	Deposit	$500.00	$1100.00
28 May	Cash withdrawal	$430.00	$670.00
31 May	Final balance		$670.00

Minimum monthly balance

From the account information given above we can see that during April the lowest balance was $250.00 and the lowest balance during May was $600.00. These lowest amounts are used to calculate the interest for each month in this *minimum monthly balance* method for determining interest.

For the 30 days of April, with a minimum balance of $250.00 and an interest rate of 6% per annum:

Interest for April $= \$250 \times \dfrac{0.06}{365} \times 30 = \$1.23$ (rounded down)

For the 31 days of May, with a minimum balance of $600.00 and an interest rate of 6% per annum:

Interest for May $= \$600 \times \dfrac{0.06}{365} \times 31 = \$3.05$ (rounded down)

Note

Rounding policy will depend on the particular rules for an account and would be decided upon by the bank or financial institution running the account.

Daily balance

In this method interest is calculated on the account balance of each day. For the account details given above the account balance is $250.00 on 1, 2, 3, 4, 5 and 6 April (i.e. 6 days), the balance changing to $750 on 7 April.

Balance of $250.00 for 6 days:

1 2 3 4 5 6 | 7 8 9 10 11 12 13 14 15 16 17 18 19 20 21 22 23 24 25 26 27 28 29 30

Interest $= \$250.00 \times \dfrac{0.06}{365} \times 6 = \$0.24$ (rounded down)

Balance of $750.00 for 4 days:

1 2 3 4 5 6 | 7 8 9 10 | 11 12 13 14 15 16 17 18 19 20 21 22 23 24 25 26 27 28 29 30

Interest $= \$750.00 \times \dfrac{0.06}{365} \times 4 = \$0.49$ (rounded down)

ISBN 9780170390194

Balance of $400.00 for 10 days:

1 2 3 4 5 6 7 8 9 10 | 11 12 13 14 15 16 17 18 19 20 | 21 22 23 24 25 26 27 28 29 30

Interest = $400.00 \times \dfrac{0.06}{365} \times 10 = \$0.65$ (rounded down)

Balance of $900.00 for 10 days:

1 2 3 4 5 6 7 8 9 10 11 12 13 14 15 16 17 18 19 20 | 21 22 23 24 25 26 27 28 29 30 |

Interest = $900.00 \times \dfrac{0.06}{365} \times 10 = \$1.47$ (rounded down)

Total interest for April = $0.24 + $0.49 + $0.65 + $1.47
= $2.85

> **Note**
>
> Again, rounding policy will depend on the particular rules for an account. In the above calculations for the April interest, if all rounding was left to the end the total would be $2.87 if rounded down, or $2.88 to the nearest cent.

Similarly, to calculate the total interest for May:

Balance of $900.00 for 4 days:

| 1 2 3 4 | 5 6 7 8 9 10 11 12 13 14 15 16 17 18 19 20 21 22 23 24 25 26 27 28 29 30 31

Interest = $900.00 \times \dfrac{0.06}{365} \times 4 = \$0.59$ (rounded down)

Balance of $1400.00 for 8 days:

1 2 3 4 | 5 6 7 8 9 10 11 12 | 13 14 15 16 17 18 19 20 21 22 23 24 25 26 27 28 29 30 31

Interest = $1400.00 \times \dfrac{0.06}{365} \times 8 = \$1.84$ (rounded down)

Balance of $600.00 for 6 days:

1 2 3 4 5 6 7 8 9 10 11 12 | 13 14 15 16 17 18 | 19 20 21 22 23 24 25 26 27 28 29 30 31

Interest = $600.00 \times \dfrac{0.06}{365} \times 6 = \$0.59$ (rounded down)

Balance of $1100.00 for 9 days:

1 2 3 4 5 6 7 8 9 10 11 12 13 14 15 16 17 18 | 19 20 21 22 23 24 25 26 27 | 28 29 30 31

Interest = $1100.00 \times \dfrac{0.06}{365} \times 9 = \$1.62$ (rounded down)

Balance of $670.00 for 4 days:

1 2 3 4 5 6 7 8 9 10 11 12 13 14 15 16 17 18 19 20 21 22 23 24 25 26 27 | 28 29 30 31 |

Interest = $670.00 \times \dfrac{0.06}{365} \times 4 = \$0.44$ (rounded down)

Total interest for May = $0.59 + $1.84 + $0.59 + $1.62 + $0.44
= $5.08

Exercise 3B
Minimum monthly balance

1 The account statement shown is for the months of August (31 days), September (30 days) and October (31 days) for an account that pays interest of 6.2% per annum calculated monthly and based on the minimum monthly balance. Interest earned in these months is added to the account on 1 January.

Calculate the interest for each of August, September and October, rounding your answer down to whole numbers of cents in each case.

Opening balance on 1 August			$245.56
Date	Deposit	Withdrawal	Balance
05 August	$300.00		$545.56
21 August		$120.00	$425.56
03 September		$65.65	$359.91
10 September	$450.00		$809.91
25 September	$1250.00		$2059.91
06 October		$750.00	$1309.91
16 October	$125.00		$1434.91
Closing balance on 31 October			$1434.91

2 The statement shows the transactions that occur in an account that pays interest of 7.45% per annum calculated monthly and based on the minimum monthly balance. Interest earned in March (31 days), April (30 days) and May (31 days) will be added to the account on 1 July.

Calculate the interest for each of March, April and May, rounding to the nearest cent.

Date	Credit	Debit	Balance
15 February		$352.65	$1256.89
07 March	$2250.00		$3506.89
03 April		$1784.25	$1722.64
29 April	$2548.00		$4270.64
07 June		$3510.00	$760.64

3 The bank statement below shows all of the activity and balances for an account from 1 July one year until 30 June of the following year. Interest is calculated monthly by applying one twelfth of the annual interest rate of 2.5% to the month's minimum balance and rounding the answer to the nearest cent. The annual interest is then added on 1 July.

Calculate the interest earned for each of the twelve months from July to June.

Date	Activity	Amount	Credit	Debit	Balance
01 Jul	Interest	$23.45	✓		$12 354.78
23 Oct	Cash deposit	$500.00	✓		$12 854.78
17 Mar	Cheque withdrawal	$7354.60		✓	$5 500.18
08 Apr	Cash withdrawal	$1000.00		✓	$4 500.18
26 Apr	Cheque withdrawal	$2780.00		✓	$1 720.18
07 May	EFT deposit	$7562.75	✓		$9 282.93

ISBN 9780170390194

Daily balance

4 The table shows the transactions and the balances for a bank account for the months of April and May.

Let us suppose that interest is paid on this account and the interest for April, May and June will be added on the first day of July.

If interest is calculated at the rate of 4% per annum based on the daily balance how much interest would the account earn for the month of April and for the month of May? (Calculate intermediate values rounded to 4 decimal places until totalling the monthly interest, then round this monthly total to the nearest cent.)

Date	Details	Amount	Balance
01 April	Opening balance		$1352.68
11 April	Deposit	$750.00	$2102.68
22 April	Chq withdrawal	$265.75	$1836.93
25 April	Deposit	$750.00	$2586.93
30 April	Final balance		$2586.93
01 May	Opening balance		$2586.93
09 May	Deposit	$750.00	$3336.93
11 May	Cash withdrawal	$675.00	$2661.93
22 May	Chq deposit	$375.25	$3037.18
23 May	Deposit	$750.00	$3787.18
25 May	Cash withdrawal	$430.00	$3357.18
31 May	Final balance		$3357.18

5 The table shows the transactions that occur in an account that pays interest of 3.25% per annum calculated on the daily balance. Interest earned for the first six months of the year is added on 1 July. Also shown is the balance on 1 January.

Calculate the interest earned for the period 1 January to 31 March calculating intermediate values rounded to 4 decimal places and then giving the total rounded to the nearest cent. (Assume the year involved is not a leap year.)

Date	Credit	Debit	Balance
01 Jan	–	–	$1256.43
23 Jan	$1250.00		
03 Feb		$356.54	
14 Feb	$876.50		
07 Mar		$782.64	
23 Mar		$1254.80	
14 Apr	$525.90		

Borrowing money

So far this chapter has been involved with earning interest on invested money. However not everyone has money to invest. Sometimes we need to borrow money from a bank or financial institution and in such cases we are charged interest on the borrowed money.

Banks make profits by charging a higher interest rate on the money they lend out than they pay on money that is invested with them.

The remainder of this chapter considers simple interest charged on loans, rather than interest paid out on invested funds.

EXAMPLE 6

How much needs to be repaid after three years to repay a loan of $8000 at 5.4% per annum simple interest?

Solution

Using $I = PRT$

$$\text{Interest} = \$8000 \times 0.054 \times 3$$
$$= \$1296.$$

Hence the amount owed $= \$8000 + \$1296$
$$= \$9296$$

$9296 needs to be repaid after three years.

Exercise 3C

Simple interest-buying
on terms

1 How much interest is charged on a loan of $25 000 borrowed for 2 years with simple interest charged at 7% per annum?

2 How much interest is charged on a loan of $8750 borrowed for 5 years with simple interest charged at 8.4% per annum?

3 How much interest is charged on a loan of $2500 borrowed for 17 months with simple interest charged at 4.75% per annum?

4 How much interest is charged on a loan of $6500 borrowed for 150 days with simple interest charged at 18% per annum?

5 How much needs to be repaid after 4 years to repay a loan of $7500 at 8.2% per annum simple interest?

6 How much needs to be repaid after five years to repay a loan of $14 500 at 12.5% per annum simple interest?

7 How much is owed after three years if $250 is borrowed at 18% per annum simple interest with no repayments being made before the end of the three years?

8 Tarni borrows $7600 to purchase a car. She agrees to repay the loan in full, plus simple interest charged at the rate of 8.5% per annum, after three years. How much will she have to pay at the end of the three years to clear the loan?

9 Frank borrows $8000 for three years with the interest fixed at 10% per annum simple interest for the three years. One year into this loan he finds he needs to borrow a further $2500 but now the simple interest rate is 12.5%. How much does he owe altogether on these two loans at the end of the three years given that he makes no repayments during the three years?

10 Ali borrows $5000 at a simple interest charge of 8% per annum. After 21 months he renegotiates the loan by paying off $3000 and having the remaining principal plus interest moved into a new loan account that charges simple interest at the rate of 7.5% per annum. If he wishes to pay off what he owes on this new account two years later how much will he have to pay to do so?

Miscellaneous exercise three

This miscellaneous exercise may include questions involving the work of this chapter, the work of any previous chapters, and the ideas mentioned in the Preliminary work at the beginning of the book.

1 Round each of the following to 2 decimal places.

  a 35.327   **b** 25.824   **c** 56.971 93   **d** 27.259 812

2 Copy and complete the following table:

	What it cost	What it was sold for	Profit as percentage of cost
a	$200	$250	
b	$450		20%
c		$2310	40%

	What it cost	What it was sold for	Loss as percentage of cost
d	$200	$190	
e	$8500		2%
f		$93 600	40%

3 a Find 25% of $500.

  b Increase $500 by 25%.

  c $500 is 25% of an amount. Find the amount.

  d After an increase of 25% an amount becomes $500. What was the amount before the increase?

4 State the 'odd one out' A, B or C, in each of the following.

  a A: 0.2   B: $\dfrac{1}{2}$ C: 20%

  b A: 2.5%   B: $\dfrac{1}{4}$ C: 0.025

  c A: 0.34   B: 75%   C: $\dfrac{3}{4}$

  d A: 0.05   B: 5%   C: $\dfrac{1}{25}$

5 What is the value at the end of five years of an initial investment of $5800 in an account paying a simple interest rate of 4.4% per annum?

6 Which pays more interest?

  A: $5000 invested for 4 years in an account paying 7% per annum simple interest, or

  B: $7000 invested for 5 years in an account paying 4% per annum simple interest.

7 $A = \pi r s + \pi r^2$

  Find A given $r = 5$ and $s = 8$. (Answer correct to one decimal place.)

8 Motor accident investigators may measure the length of a skid mark left by a braking vehicle. If a vehicle skids s metres in coming to rest then the formula

$$v = 4\sqrt{10s}$$

can be used to give an estimate for v, the speed of the vehicle in km/h at the commencement of the skid.

a Estimate the speed of a vehicle that left a skid mark of 22 metres in coming to rest.

b Crash investigators measure a skid to be approximately 50 metres. The driver of the vehicle causing this skid mark claimed that he had not been exceeding the speed limit of 110 km/h. Does the formula support his claim or not?

9 The pie graph below shows how the approximately 47 000 people employed in the resources sector in Western Australia in a particular year were distributed across the commodities.

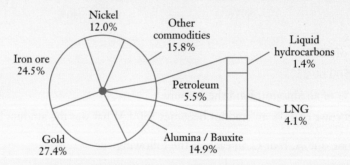

[Data source: The Chamber of Minerals & Energy.]

a Notice that the pie graph percentages add up to 100.1%:

$$24.5 + 12.0 + 15.8 + 5.5 + 14.9 + 27.4 = 100.1$$

Surely the total should be 100%. Explain how the 'extra' 0.1% could occur.

b How many of the people employed in the resources sector in Western Australia for this particular year were employed in

i Gold?　　　　　　　　　　　　　　　　**ii** LNG?

c Twelve years later the number of people employed in the resources sector in Western Australia had risen to approximately 70 000. Predict the number employed in the nickel sector at this later time, explaining any assumptions you are making.

10 A secondhand car salesperson purchases a fleet of 10 used cars for $12 000 each. He manages to sell 7 of them at a profit of 16% each. He wishes to sell the remaining 3 cars for the same price as each other and at a price that will give him a 15% profit overall for the ten cars. What price should he sell each of the remaining cars for?

4.

Compound interest

Compound interest

The previous chapter commenced by asking how much interest would be earned after 1 year, 2 years or perhaps just 6 months, if $500 were to be invested in an account paying interest at the rate of 10% per annum.

Considering a *simple interest* approach we saw that:

$$\text{Value after 2 years} = \$500 + 2 \times 10\% \text{ of } \$500$$
$$= \$500 + 2 \times \$50$$
$$= \$600$$

With *compound interest* the interest is added to the account at the end of each *compounding period*. This interest will then itself earn interest in the second and subsequent compounding periods. This 'interest on the interest' is the distinguishing feature of compound interest over simple interest.

$$\text{Value after 1 year} = \$500 + 10\% \text{ of } \$500$$
$$= \$500 + \$50$$
$$= \$550 \, (= \$500 \times 1.1)$$

$$\text{Value after 2 years} = \$550 + 10\% \text{ of } \$550$$
$$= \$550 + \$55$$
$$= \$605 \, (= \$500 \times 1.1 \times 1.1)$$
$$(\text{i.e. } \$500 \times 1.1^2)$$

The value at the end of each year is related to that of the previous year as follows:

$$\text{Value at end of one year} = \text{Value at start of that year} \times 1.1$$

If we want the value after n years:

$$\text{Value after } n \text{ years} = \text{Initial value} \times 1.1^n$$

If you managed to create the spreadsheet for simple interest mentioned in the previous chapter try to include the compound interest situation, as shown below.

	A	B	C	D	E	F	G	H
1	Amount invested ($)		$500.00					
2	Annual interest rate		10.00%					
3		Simple interest account				Compound interest account		
4	End of year	1	$550.00			1	$550.00	
5		2	$600.00			2	$605.00	
6		3	$650.00			3	$665.50	
7		4	$700.00			4	$732.05	
8		5	$750.00			5	$805.26	
9		6	$800.00			6	$885.78	
10		7	$850.00			7	$974.36	
11		8	$900.00			8	$1071.79	

In the above situation compare the value after 8 years of simple interest,

$$\text{i.e. } \$500 + 8 \text{ lots of } \$50 = \$900,$$

to the value after 8 years of compound interest,

$$\text{i.e. } \$500 \times 1.1^8 = \$1071.79.$$

- Compare the same investment after 50 years of simple interest with the investment after 50 years of compound interest.

The data for each situation is shown graphed below. Notice that the simple interest situation shows a straight line or *linear* relationship whilst the compound interest situation shows what we call *exponential* growth.

$500 invested at 10% p.a. simple interest

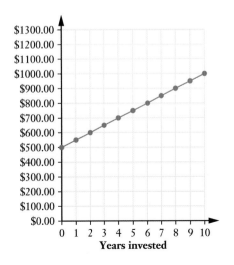

Years invested

$500 invested at 10% p.a. compound interest, compounded annually

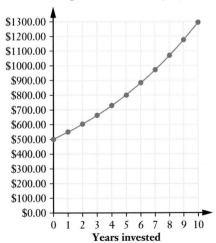

Years invested

ISBN 9780170390194

If $2500 is invested at 8% per annum compound interest, with the interest compounded annually, how much is the investment worth after 4 years?

Solution

Value at end of 1st year	= $2500 × 1.08	= $2700
Value at end of 2nd year	= $2700 × 1.08	= $2916
Value at end of 3rd year	= $2916 × 1.08	= $3149.28
Value at end of 4th year	= $3149.28 × 1.08	= $3401.22 (nearest cent)
Or, using indices:	$2500 × 1.08^4	= $3401.22 (nearest cent)

The value of the investment after 4 years is $3401.22, to the nearest cent.

Alternatively we could use the built-in capability of some calculators, or some internet programs, to perform compound interest calculations.

Explore such capabilities.

If $12 000 is invested at 6.2% per annum compound interest, with interest compounded annually, how much is the investment worth after 10 years?

Compare the final value with that of the same amount invested for the same time but with the 6.2% per annum calculated as simple interest.

Solution

Value at end of 10th year $= \$12\,000 \times 1.062^{10}$

$\qquad\qquad\qquad\qquad = \$21\,899.11$ (to the nearest cent)

The value of the investment after 10 years is $21 899.11, to the nearest cent.

With simple interest, interest after 10 years $= \$12\,000 \times 0.062 \times 10$

$\qquad\qquad\qquad\qquad\qquad\qquad = \$7440$

Hence with simple interest the final value $\quad = \$12\,000 + \$7440$

$\qquad\qquad\qquad\qquad\qquad\qquad\qquad = \$19\,440$

After ten years, the compound interest method returns $2459.11 more than the simple interest method.

Other compounding periods

In the compound interest situations encountered so far the interest has been compounded *annually*. Suppose that the 8% per annum of example 1 above were compounded every six months instead of every 12 months. The 8% per annum would now mean 4% per compounding period and in the 4 years there would be eight compounding periods. In this case:

$$\text{Value of investment at end of 4th year} = \$2500 \times 1.04^8$$
$$= \$3421.42.$$

The six-monthly compounding has returned $20.20 more interest after four years than the annual compounding did.

Simple and compound interest

Spreadsheets – Simple and compound interest

Compound interest table

EXAMPLE 3

Determine the interest paid on $5000 invested at 8% per annum compound interest for 3 years, with the interest compounded

a   annually   **b**   every six months   **c**   quarterly.

Solution

a   Value of investment after 3 years $= \$5000 \times 1.08^3$

$$= \$6298.56$$

Thus the interest paid $= \$6298.56 - \$5000$

$$= \$1298.56$$

b   If interest is compounded every 6 months the 8% per annum means 4% each compounding period.

In 3 years there will be 6 compounding periods.

Value of investment after 3 years $= \$5000 \times 1.04^6$

$$= \$6326.60 \text{ (to nearest cent)}$$

Thus the interest paid $= \$6326.60 - \$5000$

$$= \$1326.60$$

c   If interest is compounded quarterly (every 3 months) the 8% per annum means 2% each compounding period.

In 3 years there will be 12 compounding periods.

Value of investment after 3 years $= \$5000 \times 1.02^{12}$

$$= \$6341.21 \text{ (to nearest cent)}$$

Thus the interest paid $= \$6341.21 - \$5000$

$$= \$1341.21$$

Borrowing money

As mentioned in the chapter on simple interest, it is not always about investing money – sometimes we have to borrow money *from* a financial institution rather than lending money *to* it. We then have to pay interest to the institution for the money we have borrowed.

Suppose we borrow $1000 for six months and the interest rate we are charged is 12% per annum.

Now 12% of $1000 is $120 so, using a simple interest approach, the total interest we would be charged for the six months is $60.

If we use a monthly compound interest approach then the amount owed at the end of the 6 months is

$$\$1000 \times 1.01^6.$$

Thus $1061.52 is owed, i.e. interest of $61.52.

EXAMPLE 4

How much will be owed after 3 years on a loan of $4000 with compound interest charged at 8% per annum, compounded daily?

Solution

In 3 years, there will be 3×365 (= 1095) compounding periods.

$$\text{Value of investment after 3 years} = \$4000 \times \left(1 + \frac{0.08}{365}\right)^{1095}$$
$$= 5084.86, \text{ to the nearest cent}$$

Again explore the capability of your calculator, and of some internet programs, to perform calculations involving compound interest, especially when compounding occurs monthly or weekly.

Exercise 4A

1 If $5000 is invested at 5% per annum compound interest with the interest compounded annually, how much is the investment worth after 4 years?

2 If $200 is invested at 8% per annum compound interest with the interest compounded annually, how much is the investment worth after 25 years?

3 How much is owed after three years if $250 is borrowed at 18% per annum compound interest with no repayments being made before the end of the three years and interest is compounded annually?

4 Determine the interest paid on $1000 invested at 4% per annum compound interest for 3 years, with the interest compounded

   a   annually   **b**   every six months   **c**   quarterly.

5 Determine the interest paid on $5000 invested at 12% per annum compound interest for 3 years, with the interest compounded

   a   annually   **b**   every six months   **c**   monthly.

6 How much is owed after two years if $2000 is borrowed at 6% per annum compound interest with interest compounded monthly and no repayments being made before the end of the two years?

7 Determine the value of an investment of $2000 after 2 years if the interest rate is 12% per year calculated

    a as simple interest

    b as compound interest compounded annually

    c as compound interest compounded monthly

    d as compound interest compounded daily.

8 A product called a 'reverse mortgage' allows older people who own their own home to take out a loan to help their financial situation whilst on a pension. The amount that needs to be repaid increases with time as the interest payments are added but no repayments are made until the person taking out the loan dies. The home is then sold, the loan plus interest is repaid from the proceeds and any remaining funds are distributed according to the will of the deceased person.

An elderly couple borrow $40 000 with interest compounded monthly at a rate of 9% per annum. How much will need to be repaid on this loan 15 years later?

9 Copy and complete the following table to compare various forms of interest for a loan of $10 000 at 8% per annum. (You may wish to use a spreadsheet on a computer or calculator.)

	$10 000 borrowed at 8% per annum			
	Simple interest	Compounded annually	Compounded every 6 months	Compounded quarterly
Initial amount borrowed				
Amount owed after 1 year				
Amount owed after 2 years				
Amount owed after 3 years				
Amount owed after 4 years				
Amount owed after 10 years				
Amount owed after 20 years				

10 Copy and complete the following table to compare various forms of interest for an investment of $2000 at 12% per annum. (You may wish to use a spreadsheet on a computer or calculator.)

	$2000 invested at 12% per annum			
	Simple interest	Compounded annually	Compounded every 6 months	Compounded monthly
Initial balance				
Balance after 1 year				
Balance after 2 years				
Balance after 3 years				
Balance after 4 years				
Balance after 10 years				
Balance after 20 years				

ISBN 9780170390194

Inflation and depreciation

Inflation was mentioned in chapter 2 as an example of the everyday use of percentages. It is again mentioned in this chapter (along with the concept of depreciation) because, whilst the concepts are not examples of compound interest, they each similarly involve the repeated multiplication by a number. For example repeated multiplication by 1.1 for a 10% inflation rate, or by 0.9 for a 10% depreciation rate.

Consider again the 'reversible mortgage' situation explained in question 8 of the previous exercise:

> *A product called a 'reverse mortgage' allows older people who own their own home to take out a loan to help their financial situation whilst on a pension. The amount that needs to be repaid increases with time as the interest payments are added but no repayments are made until the person taking out the loan dies. The home is then sold, the loan plus interest is repaid from the proceeds and any remaining funds are distributed according to the will of the deceased person.*

The elderly couple may be alarmed that their initial loan of $40 000 'blows out' to more than $150 000 ($40 000 \times 1.0075^{180}$) after 15 years because they are allowing the interest to accumulate rather than reducing it by making repayments. However in this fifteen years the house that will be sold to pay off the loan will have **appreciated** (increased) in value due to **inflation** (the general rise in prices with time). Suppose their house was worth $310 000 at the start of the fifteen years and that the annual inflation rate is 3.4%.

$$\text{Value of the house after 15 years} = \$310\,000 \times 1.034^{15}$$
$$= \$512\,000 \text{ to the nearest } \$1000.$$

Hence whilst the loan has increased by approximately $110 000 in the fifteen years the asset that will be used to pay off the loan has increased by approximately $202 000 in the same time due to inflation.

- Consider this reverse mortgage situation with the $40 000 borrowed at 12% per annum compounded monthly for 25 years, and house values rising at just 3% per annum.

EXAMPLE 5

A particular item has a current value of $8600, this value rising in line with inflation.

If we assume a constant annual inflation rate of 4.2% what will the value of this item be

a   2 years from now?   b   25 years from now?

Solution

With an annual inflation rate of 4.2% we have annual multiplication by 1.042.

a   Value of the item 2 years from now $= \$8600 \times 1.042^2$
$$= \$9337.57 \text{ to the nearest cent.}$$

Assuming a constant annual inflation rate of 4.2% the item will have a value of approximately $9300 two years from now.

b   Value of the item 25 years from now $= \$8600 \times 1.042^{25}$
$$= \$24\,054.23 \text{ to the nearest cent.}$$

Assuming a constant annual inflation rate of 4.2% the item will have a value of approximately $24 000 twenty-five years from now.

Not everything increases in value. Some items fall in value as they get older. They are said to **depreciate**. A car for example will fall in value as it gets older (until it becomes so old that it becomes rare and collectible in which case its value could start to rise).

pixabay.com/lookapic

EXAMPLE 6

A particular car has a new value of $28 000. If we assume a constant annual depreciation rate of 7% what will the value of this car be when it is

a 2 years old?

b 8 years old?

Solution

With a constant annual depreciation rate of 7% we have annual multiplication by 0.93.

a Value of the car when 2 years old = $28 000 \times 0.93^2$
$$= \$24\,217.20 \text{ to the nearest cent.}$$

Assuming a constant annual depreciation rate of 7% the car will have a value of approximately $24 200 when it is two years old.

b Value of the car when 8 years old = $28 000 \times 0.93^8$
$$= \$15\,668.29 \text{ to the nearest cent.}$$

Assuming a constant annual depreciation rate of 7% the car will have a value of approximately $15 700 when it is eight years old.

Exercise 4B

1 A particular car has a new value of $32 000. If we assume a constant annual depreciation rate of 12% what will the value of the car be when it is

a 1 year old?

b 5 years old?

2 A particular house has a current value of $350 000. If we assume a constant annual inflation rate of 4.8% what will the value of this house be

a 2 years from now?

b 20 years from now?

c 50 years from now?

3 Assuming a constant annual inflation rate of 4% what will be the cost of a particular type of chocolate bar in 20 years time if it costs $2.20 now?

Suppose instead that the annual inflation rate for the period was 8% rather than 4%. What would the chocolate bar cost in 20 years time now?

4 A car has a current value of $32 000. If we assume a constant depreciation rate of 7.2% per year what will the value of this car be

a 3 years from now?

b 5 years from now?

c 10 years from now?

5 The increased availability of a particular commodity causes its price per kg to depreciate by 5.2% each year for a five year period. At the start of this period the commodity cost $135 per kg. What was the cost per kg of this commodity at the end of the five year period?

ISBN 9780170390194

Miscellaneous exercise four

This miscellaneous exercise may include questions involving the work of this chapter, the work of any previous chapters, and the ideas mentioned in the Preliminary work at the beginning of the book.

1 a Find 20% of $40 **b** Find 40% of $20

   c Find 30% of $120 **d** Find 5.8% of $120

   e Increase $750 by 16% **f** Decrease $430 by 18%.

2 The number of years, T, required for an investment of P, earning simple interest at a rate of R% per annum, to earn I interest can be found using the formula:

$$T = \frac{100I}{PR}$$

Find T if

   a $P = 300, R = 5, I = 90$ **b** $P = 540, R = 7.5, I = 324$

   c $P = 75.80, R = 5, I = 37.90$ **d** $P = 17\,500, R = 4, P + I = 19\,950$

3 If inflation is running at a steady 4.4% per annum, and considering inflation to be the only reason for prices to rise, what will be the cost of an item in ten years if its cost now is $240?

4 Anje wishes to invest $8000 for 3 years. She considers three schemes:

   A: Simple interest at 9.00% per annum.

   B: Compound interest at 8.15% per annum compounded quarterly.

   C: Compound interest at 8.08% per annum compounded monthly.

Which scheme should she choose to maximise the value of the account at the end of the three year period and what would that maximum value be?

5 An elderly couple who own their own home decide to borrow $60\,000.

The loan, plus interest, is to be paid off from the proceeds of the sale of the house upon the death of the last surviving member of the couple (or earlier if the couple decide to sell the house earlier).

Interest is to be compounded annually and the interest rate is fixed at 8.5% per annum.

The table below shows the interest to be added and the amount owing at the end of a number of years.

Year	Interest for the year	Loan amount
1	$5100	$65\,100
2	$5533.50	$70\,633.50
3		
4		
10		
25		

Copy and complete the table.

6 Tax changes announced.

Suppose that the government of a country for which the income tax rates were as given in the table below left announces that as from the following tax year the rates will change to those shown in the table below right:

Current tax year	
Taxable income	**Tax rate**
$0 to $19 400	0%
$19 401 to $37 000	19%
$37 001 to $80 000	33%
$80 001 to $180 000	37%
$180 001+	45%

Next tax year	
Taxable income	**Tax rate**
$0 to $24 000	0%
$24 001 to $45 000	15%
$45 001 to $90 000	30%
$90 001 to $180 000	38%
$180 001+	46%

Suppose that you work for a newspaper and, for a proposed article about the tax changes, you are asked to produce a table like that shown below for the current tax year but your table should be for the 'next tax year' rates:

Taxable income	Income tax to pay
$0 to $19 400	Nil
$19 401 to $37 000	19% of the taxable income over $19 400
$37 001 to $80 000	$3344 + 33% of the taxable income over $37 000
$80 001 to $180 000	$17 534 + 37% of the taxable income over $80 000
$180 001+	$54 534 + 45% of the taxable income over $180 000

AND a table showing what people on various taxable incomes will save, both over a full year and per week, when the new rates are applied. i.e. a table like that shown below:

What the tax changes will mean to you		
Taxable income	**Savings per year**	**Savings per week**
$5 000		
$15 000		
$25 000		
$35 000		
$50 000		
$60 000		
$75 000		
$100 000		
$125 000		
$150 000		
$200 000		

Produce completed tables as required.

Challenge: For what taxable incomes would the new rates mean you would be paying more total income tax than under the old rates?

ISBN 9780170390194

5.

Other financial considerations

- Wages and piecework
- Salary
- Comparing prices
- Foreign currency
- Shares
- Price to earnings ratio, or P/E
- Government allowances and pensions
- Budgeting
- Miscellaneous exercise five

Situation One

Jemima Quek works for a company and her normal rate of pay is $23.40 per hour.

However:
- If she works more than 8 hours on any weekday, further hours worked that day are paid at 'time and a half'.
- Any Saturday hours are paid at 'time and a half'.
- Any Sunday hours are paid at 'double time'.

The partially completed wage slip for a particular week in the life of Jemima Quek is shown below with question marks indicating missing values.

	Total hours worked	Normal hours worked	Time and a half hours worked	Double time hours worked	Earnings
Wage slip for: JEMIMA QUEK					
		$23.40 / hour	?	$46.80 / hour	
Mon	8	8	0	0	?
Tues	8	8	0	0	?
Wed	9	8	1	0	?
Thur	10	?	?	?	?
Fri	8	?	?	?	?
Sat	4	?	?	?	?
Sun	4	?	?	?	?
				Total earnings for the week:	?

Copy and complete the wage slip.

Situation Two

Tom Jacowitz is looking in the newspaper for clerical jobs and sees the advertisement partially shown on the right.

Tom is 22 years old and wants to know:
- how much this job would pay him per week
- how much this job would pay him per fortnight
- how much this job would pay him per month
- how much his weekly income would increase under this scheme when he becomes 25.

Determine answers to what Tom wants to know, including with your answers brief mention of any assumptions you are making.

WANTED

ACCOUNTS CLERK

Annual salary:

Aged 20–24 $58 000

Aged 25 and over $65 000

Promotional opportunities exist thereafter.

Wages and piecework

Wage sheet

Income-commission and piecework

Earning money

If you work for an *employer* there are a number of ways the amount you are to be paid might be calculated. (One such example would be payment by *commission*, an idea considered in Chapter Two, Percentages.) Some *employees* receive a weekly or fortnightly *wage*. This is an amount paid for a particular quantity of work, for example the $23.40 for each normal hour worked by Jemima Quek in Situation One on the previous page. The employee may be able to earn a higher rate of pay by working *overtime*, again as in Situation One.

Alternatively the 'particular quantity of work' might be a piece of work rather than an hour of work. For example, a carpenter could be paid $50 for each chair frame they complete, a fruit picker could be paid a certain amount for each carton of fruit they pick. This is called *piecework* (payment by the 'piece').

Salary

Alternatively an employee may be on a *salary* which is a fixed amount earned per year. For example the $58 000 salary offered in the advertisement mentioned in Situation Two on the previous page. This might be paid as a regular weekly amount (equal to the annual salary divided by 52), more commonly as a fortnightly amount (equal to the annual salary divided by 26), or perhaps as a monthly amount (equal to the annual salary divided by 12).

Exercise 5A

Each of questions 1 to 6 involve situations in which the following payment rules apply.

- The first 8 hours worked on any week day are paid at the normal hourly rate.

- If the employee works more than the basic eight hour day on any day then the next two hours that day are paid at 'time and a half' and any hours after these extra two on that day are paid at 'double time'.

- Any Saturday hours are paid at 'time and a half'.

- Any Sunday hours are paid at 'double time'.

Calculate the amount each person earns for the week shown.

1

NAME: Jamie Clark	
Normal rate: $18.50/hr	
Mon	8 hours
Tue	8 hours
Wed	8 hours
Thur	8 hours
Fri	8 hours
Sat	–
Sun	–

2

NAME: Ben Carto	
Normal rate: $26.40/ hr	
Mon	8 hours
Tue	10 hours
Wed	8 hours
Thur	11 hours
Fri	8 hours
Sat	–
Sun	–

3

NAME: Jen Lee	
Normal rate: $22.40/hr	
Mon	8 hours
Tue	11 hours
Wed	8 hours
Thur	11 hours
Fri	8 hours
Sat	5 hours
Sun	–

4

NAME: Amrit Poller	
Normal rate: $32.24/ hr	
Mon	8 hours
Tue	8 hours
Wed	8 hours
Thur	10 hours
Fri	11 hours
Sat	4 hours
Sun	4 hours

5

Naomi Clankart	
Normal rate of pay: $17.60 per hour	
Monday	0800 to 1700 with one hour off for lunch (unpaid).
Tuesday	0800 to 1800 with one hour off for lunch (unpaid).
Wednesday	0730 to 1730 with one hour off for lunch (unpaid).
Thursday	0800 to 1900 with one hour off for lunch (unpaid).
Friday	0800 to 1700 with one hour off for lunch (unpaid).
Saturday	–
Sunday	–

6

Mai Lee Wong	
Normal rate of pay: $29.80 per hour	
Monday	0730 to 1600 with 30 minutes off for breaks (unpaid).
Tuesday	0730 to 1600 with 30 minutes off for breaks (unpaid).
Wednesday	0730 to 1700 with 30 minutes off for breaks (unpaid).
Thursday	0730 to 1700 with 30 minutes off for breaks (unpaid).
Friday	0730 to 1930 with 1 hour off for breaks (unpaid).
Saturday	0900 to 1200 with no breaks.
Sunday	0900 to 1200 with no breaks.

7 Bob works one day a week for a bike shop, servicing bikes. He is paid $35 for each bike that he services.

In one month he services 25 bikes.

How much will Bob get paid for this work?

Alamy Stock Photo/Ken Gillespie Photography

8 Nigel picks plums and is paid 45 cents for each kilogram picked.

How much will Nigel be paid for picking

a   50 kg?   **b**   150 kg?

9 Tom is paid $75 for each single bed frame he makes, assembles then unassembles and packs ready for sending and $95 for each double bed frame that he similarly makes, assembles, unassembles and packs.

In one week he carries out these tasks for 6 single beds and 8 double beds.

What does Tom earn by completing these tasks?

10 Anne works in a bedding factory as a machinist making quilts. She is paid $6.50 for each single quilt she makes, $9.50 for each double size quilt she makes and $12.00 for each queen size quilt she makes. How much will Anne earn in a week that she makes 52 single quilts, 36 double size quilts and 16 queen size quilts?

11 Sharup earns $31.40 per hour for a basic 40 hour week. He also tends to work three overtime hours per week for which he is paid 'time and a half'. He has four weeks holiday per year during which the company pays him his basic 40-hour week wage without overtime.

Sharup is offered a salaried position as a foreman for which his salary will be $75 000 per year.

He asks for your advice as to whether he should take the salaried position. What would your advice be?

12 Remember,   1 year = 52 weeks,
                              26 fortnights,
                              12 months.

a $69 000 per annum, how much per month?

b $65 000 per annum, how much per week?

c $56 000 per annum, how much per fortnight (to the nearest cent)?

d $1560 per fortnight, how much per year?

e $1290 per week, how much per year?

13 List the following in order of size, from the one that would give the greatest amount per year to the one that would give the smallest amount per year. (Assume stated pay rate continues through holidays.)

$86 000 per annum.

$3210 per fortnight.

$1680 per week.

$38.75 per hour, 40 hour week, no overtime.

$8000 per month.

$75 000 per year.

$41.20 per hour, 38 hour week, no overtime.

$7000 per month.

Comparing prices

Consider the three different size jars of *Blend Rite Coffee* shown below.

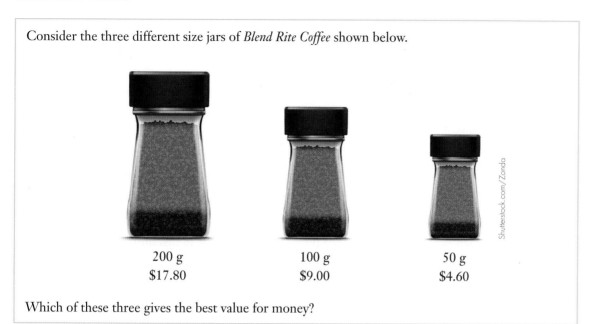

| 200 g | 100 g | 50 g |
| $17.80 | $9.00 | $4.60 |

Which of these three gives the best value for money?

Notice that in the situation shown above, if we double the quantity in the small jar (50 g) we get the quantity in the middle sized jar (100 g), and doubling again gives the quantity in the large size jar (200 g).

Did you use this 'doubling relationship' to calculate the best buy or did you use some other technique?

Try the following 'best buy' situation:

Consider the three different size jars of *Blend Smooth Coffee* shown below.

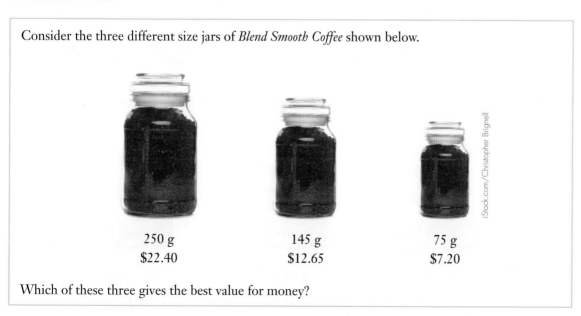

| 250 g | 145 g | 75 g |
| $22.40 | $12.65 | $7.20 |

Which of these three gives the best value for money?

The second situation on the previous page did not have the nice 'doubling relationship' between the amounts of coffee each jar held that the first situation had. Whilst this may have made a mental comparison of value more difficult we could still investigate which jar offered the best value by considering the price per gram (older British spelling is gramme), or some other convenient amount, for example price per 100 grams.

$22.40 per 250 grams $= \dfrac{\$22.40}{250}$ dollars per gram

$= 0.0896$ $/g$ (which is 8.96 $/100 g.)

$12.65 per 145 grams $= \dfrac{\$12.65}{145}$ dollars per gram

$= 0.0872$ $/g (to 4 decimal places) (which is 8.72 $/100 g.)

$7.20 per 75 grams $= \dfrac{\$7.20}{75}$ dollars per gram

$= 0.096$ $/g (which is 9.60 $/100 g.)

Comparisons can then easily be made once the prices have been given in this **unit price** form, i.e. where the price is quoted per *unit of measure*. (In the above calculations the units of measure used are 1g and 100 g.)

Note: In 2009 the Australian Federal Government introduced legislation making it compulsory for all large supermarkets to display *Unit Price Information* with particular grocery items. Look for this the next time you are in a large supermarket.

If you are good at doubling sums of money then determining the best buy of the two breakfast cereal packets below left is straightforward, but the two below right is much harder to do mentally.

500 g
$7.54

250 g
$3.90

500 g
$7.54

350 g
$5.18

However, if the unit price information is given price comparison is made much easier:

500 g
$7.54
0.01508 cents per gram

350 g
$5.18
0.0148 cents per gram

250 g
$3.90
0.00156 cents per gram

Exercise 5B

1 Write each of the following as 'prices per gram'.

   a $50 for 800 grams.   **b** $18.60 for 200 grams.

   c $8.40 for 250 grams.   **d** $16.80 for one kilogram.

2 Write each of the following as 'prices per 100 grams', giving answers rounded to two decimal places.

   a $17.80 for 300 grams.

   b $9.70 for 150 grams.

   c $17.40 for 1.35 kilograms.

   d $16.80 for one pound where one pound is an older unit of measure and is equal to 0.45359 kg.

3 Write each of the following as 'prices per kilogram'.

   a $4.80 for 200 grams.   **b** $11.60 for 500 grams.

   c $20 for 1 kilogram.   **d** $75 for 5 kilogram.

4 Express each of the following prices as dollars per litre and hence determine 'the best buy'.

3 L
$6.75

1.8 L
$4.50

2.4 L
$5.40

5 Express each of the following prices as $/100 g and hence rank the deals in order of value, best value first.

400 g
$4.65

600 g
$6.40

1 kg
$10.50

6 Rank each of the following in order of value, best value first.

375 g
$5.20

250 g
$3.95

500 g
$7.00

Foreign currency

If you travel overseas you will need to exchange Australian currency, i.e. Australian dollars, for the currency used in the country you are visiting, for example, British pounds (£), European euros (€), Japanese yen (¥), American dollars ($), etc.

Note: The word **dollar** is used for the basic unit of currency in a number of countries including Australia, USA, New Zealand, Singapore and others. If we need to distinguish between them then note that:

AUD, $A, Aus, A etc, may be used to indicate the Australian dollar.
USD, $US or US$ may be used to indicate the US dollar.
NZD, $NZ or NZ$ may be used to indicate the New Zealand dollar.
SGD, $S, $Sing, S$ etc may be used to indicate the Singapore dollar.
Etc.

You might decide to exchange some Australian dollars for another currency before you leave Australia so that you have some of the foreign currency to take with you and to use on your arrival in the other country. You may choose to use some of your Australian money to purchase bank notes in the foreign currency. Alternatively you might purchase *traveller's cheques* that can be exchanged for currency in the other country or perhaps you will choose to 'load up' a *pre-paid travel card* with foreign currency and use that to access foreign currency whilst abroad.

You might have access to foreign currency whilst overseas by using your Australian bank card in a bank in that country. This will give you some foreign currency and the equivalent amount in Australian dollars will be deducted from your Australian bank balance (quite possibly with a fee for the transfer included as well).

Foreign currency is also important if you are buying something 'on line' from an overseas company. If you use your credit card to pay for the goods the amount you pay in the foreign currency will be converted to the equivalent number of Australian dollars and that sum will be charged to your credit card account.

To determine how much foreign currency we could get for our Australian dollars we need to know the **exchange rate** for the two currencies, i.e the rate at which one currency will be exchanged for the other. These rates can vary from day to day. Current rates are often on display in banks, post offices, travel agents and offices of foreign exchange *brokers*. (A *broker* is a person or company who arranges a transaction between a buyer and a seller, usually for a fee or commission.)

The following table gives the amount one Australian dollar will buy for a number of foreign currencies.

Foreign location	Currency	Symbol	$1 Australian will buy
Britain	Pound Sterling	£	0.6572 British Pounds
Canada	Canadian Dollar	CA$	1.0293 Canadian Dollars
Europe	Euro	€	0.7704 Euros
Hong Kong	Hong Kong Dollar	HK$	7.9959 Hong Kong Dollars
India	Rupee	₹	54.9071 Indian Rupees
Indonesia	Rupiah	Rp	9990.3900 Indonesian Rupiah
Japan	Yen	¥	96.1600 Japanese Yen
Malaysia	Ringgit	RM	3.1899 Malaysian Ringgits
New Zealand	New Zealand Dollar	NZ$	1.2394 New Zealand Dollars
Singapore	Singapore Dollar	S$	1.2784 Singapore Dollars
South Africa	Rand	R	9.1989 South African Rands
Thailand	Baht	฿	30.7500 Thailand Baht
USA	American dollar	US$	1.0315 American Dollars

Note: The rates given in the above table will be used as the exchange rates for the following examples and for the exercise that follows. As mentioned earlier, these rates can change from day to day. You might like to compare the above figures, which were based on the rates at the time this page was created, with the rates applying on the day you are reading this page.

With one Australian Dollar (A$1) being equal to 54.9071 Indian Rupees (₹):

a how many Rupees could be bought for A$250?

b how many A$ could be bought for 2500 Rupees?

Solution

a   A$1.00 buys   54.9071   Rupees

∴   A$250 buys   250 × 54.9071 Rupees

                     = 13 726.775   Rupees

                     = 13 726   Rupees, (rounded down to whole Rupees).

b We need to determine 'how many lots of 54.9071 Rupees there are in 2500 Rupees'.

Now $\dfrac{2500}{54.9071} = 45.53$ (correct to 2 decimal places).

Thus 2500 Rupees will buy A$ 45.53, to the nearest cent.

Exercise 5C

1 With 1 Australian Dollar (A$1) being equal to 1.2394 New Zealand Dollars (NZ$), and rounding answers to the next whole dollar *down*:

a How many NZ$ could be bought for A$750?

b How many A$ could be bought for NZ$1250?

2 With 1 Australian Dollar (A$1) being equal to 96.16 Japanese Yen (¥):

a How many Yen could be bought for A$8000?

b How many A$ could be bought for ¥8000? (Answer to nearest cent.)

3 With 1 Australian Dollar (A$1) being equal to 9.1989 South African Rand (R):

a How many South African Rand could be bought for A$300? (Nearest Rand.)

b How many A$ could be bought for R5000? (Nearest cent.)

4 With 1 Australian Dollar (A$1) being equal to 0.6572 British Pounds (£):

a How many British Pounds could be bought for A$850? (To 2 decimal places.)

b How many A$ could be bought for £5000? (Answer to nearest dollar.)

5 With 1 Australian Dollar (A$1) being equal to 3.1899 Malaysian Ringgits (RM):

a How many Ringgits could be bought for A$2500? (To nearest Ringgit.)

b How many A$ could be bought for RM8000? (To nearest cent.)

6 Yoshi changed A$2000 to Japanese Yen (¥) to use as spending money for his short holiday in Japan. During the holiday, of this spending money, Yoshi spent ¥131 700.

Upon his return to Australia Yoshi changed the remaining Yen (rounded down to a multiple of 1000 Yen) back to Australian dollars.

Using an exchange rate of A$1 equaling ¥96.16 for both transactions, how much Australian money did Yoshi get back (rounded down to a multiple of 5 cents)?

7 Naomi changes A$1000 to Singapore Dollars (S$) to use as holiday spending money in Singapore.

Naomi spends S$1025 and upon her return to Australia Naomi changes the remaining Singapore Dollars, rounded down to a multiple of 2 Singapore Dollars, back to Australian Dollars (rounded down to a multiple of 5 cents).

Using an exchange rate of A$1 equaling S$1.2784 for both transactions, how much Australian money did Naomi get back?

8 Pete pays some Australian Dollars to get US$2000, at an exchange rate of A$1 equaling US$1.0315, in readiness for an anticipated trip to the US.

Unfortunately the trip is cancelled and so Pete changes the US$2000 back to Australian Dollars, but now the rate has changed to A$1 = US$1.0525.

How much Australian money did Pete pay out to get the US$2000 and how much Australian currency does Pete get back (give both answers to the nearest cent)?

 ISBN 9780170390194

9 Wendy returns to Australia after an around the world trip involving visits to the United States, Britain, various European countries and Hong Kong.

Upon her return she collects up the various foreign banknotes she has left and finds she has US$60, £75, €125 and HK$60.

Using the exchange rates given in the table just prior to this exercise calculate a total A$ value of this collection of foreign banknotes, rounding your total to the nearest Australian dollar.

10 Connor ordered some computing equipment from online stores based in the United States, Britain, Europe, Canada, Singapore and Australia.

In all he spent:	US$545,
and	£1250,
and	€660,
and	CA$165,
and	S$120,
and	A$480.

Using the exchange rates given in the table just prior to this exercise calculate the total value of this spending, in Australian dollars, to the nearest ten dollars.

11 When a dealer in foreign currency (a currency broker) sells you some foreign currency they make money by either charging you a fee, as in part **a** below, or by having a different 'buy rate' and 'sell rate', as in part **b** below (or perhaps by charging a fee AND having different buy and sell rates). The buy rate is the rate at which the dealer will *buy* foreign currency *from* someone and the sell rate is the rate at which the dealer will *sell* the foreign currency *to* someone.

a Tom wishes to use some Australian money to buy US bank notes to the value of US$250. The currency broker he goes to offers to sell him US$1.0315 for each A$ but also charges a fee of A$15 on top, and then rounds up to the next multiple of 5 cents.

How much Australian money will the US$250 cost Tom altogether?

b Julie returns from her holiday in America and changes her remaining US currency back to Australian dollars. She exchanges US$500 with a broker who pays her A$0.9295 for each US$1.

Later that day Leroy wants to purchase some US banknotes for his trip to America. The broker offers him the US$500 selling each US$ for A$1.0095. If Leroy buys all US$500 how much has the broker made in this 'buy from Julie and sell to Leroy' transaction?

TECHNOLOGY

There are a number of online currency calculators – investigate.

Shares

Some people choose to invest money by buying shares in a company.

If the company does well the share price may increase.

If the company makes a profit it may pay out some or all of this profit to the shareholders.

Hence the investor may see the value of their investment rise, as the share price increases, and also could receive an annual payment from the company, called a **dividend**, which is the investor's share of the company profits.

- The company may decide not to distribute all of the profits to the investors, preferring to re-invest some back into the company, allowing the company to expand and increase future profits.

- The company may not make a profit every year. Some years it may just 'break even' or perhaps make a loss.

- The share price could drop in value, decreasing the total value of the shares owned.

All of the shares an investor owns in various companies make up the investor's share **portfolio**.

EXAMPLE 2

An investor owns shares in four companies as shown by the portfolio details below, which also shows the dividend per share for each holding.

Company	Number of shares	Value per share	Dividend per share
BCD Limited	5000	$0.86	$0.05
EFG Group	540	$23.50	$1.45
HIJ Resources	2500	$3.40	$0.28
KLM Group	7000	$0.54	No dividend

Calculate the total value of this portfolio, based on the share values given, and the total dividend the investor will receive.

Solution

Value of BCD Ltd shares:	5000 shares at $0.86 each	=	$4 300
Value of EFG Group shares:	540 shares at $23.50 each	=	$12 690
Value of HIJ Resources shares:	2500 shares at $3.40 each	=	$8 500
Value of KLM Group shares:	7000 shares at $0.54 each	=	$3 780
	Total value	=	$29 270

Dividend from BCD Ltd shares:	$5000 \times \$0.05$	=	$250
Dividend from EFG Group shares:	$540 \times \$1.45$	=	$783
Dividend from HIJ Resources shares:	$2500 \times \$0.28$	=	$700
Dividend from KLM Group shares:	$7000 \times \$0.00$	=	$0
	Total dividend	=	$1 733

Price-to-earnings ratio, or P/E

If a share with a value of $20 paid a dividend of $4 in the most recent 12 month period we say that its **price-to-earnings ratio**, or P/E, is $\frac{20}{4}$, i.e. 5.

Similarly a share with a value of $32 paying a dividend of $2 in the most recent 12 month period has a price-to-earnings ratio of 16 $(= \frac{32}{2})$.

$$\text{Price to earnings ratio, or P/E,} = \frac{\text{Share price}}{\text{Earnings in past year}}.$$

The P/E tells us how many dollars worth of share value we must pay to get one dollar of earnings in the year. I.e. in the first case above, each $5 of share value earns us $1 of dividend and in the second case each $16 of share value earns us $1 of dividend.

We can also view the P/E of a share as the number of years it will take for the dividend to pay for the share. In the first case above the P/E of 5 means that 5 years of continued dividend will pay for the share whereas in the second example, with a P/E of 16, it will take 16 years of continued dividends to pay for the share.

So does a lower P/E indicate a better share?

Oh if only picking the right shares to invest in were that simple. We could all make a fortune!

Let us consider two shares, one with a P/E of 10 and the other with a P/E of 8. For the first one we need to spend $10 to get earnings of $1 whereas for the second one we need to spend only $8 to get earnings of $1. This would suggest that the second share, with its lower P/E, is the better buy, but this might not be the case. The first share might be in a company that has excellent prospects for big profits in the future, but is not yet in a position to deliver those anticipated profits. This could make shares in this company very popular and investors could be prepared to pay a higher price for a share that promises good returns in the future. This will drive the price up but with the higher profits still to come the current P/E could be high.

So a high P/E may indicate shares that are over priced but it could also indicate a share for which there are high hopes of large profits in the future.

The price to earnings ratio is just one piece of useful information. It does allow comparison between shares to be made, and monitoring it over a period of time allows us to consider changes, but it is just one piece of information. Future prospects, share price history, earnings history, company expectations for the future and other aspects all need to be considered too before we can decide which shares are best to buy – and even then we might get it wrong!

Note: If a company makes either no profit or a negative profit (i.e. a loss) we tend to say that the price earnings ratio is not applicable (because there are no earnings).

EXAMPLE 3

Determine the price to earnings ratio for a share with a price of $5.50 and dividends in the last twelve months totaling 44 cents per share.

Solution

$$\text{Price-to-earnings ratio} = \frac{\text{Share price}}{\text{Earnings in past year}}$$

$$= \frac{\$5.50}{\$0.44}$$

$$= 12.5$$

The price to earnings ratio for this share is 12.5.

Exercise 5D

For each of questions 1 to 4 find both the total value and the total dividend due for each of the given share portfolios.

1 **Portfolio One**

Company	Number of shares	Value per share	Dividend per share
RK Industries	3500	$8.74	$0.84
DG Resources	7500	$2.54	$0.45
JA Corporation	800	$15.45	$0.88

2 **Portfolio Two**

Company	Number of shares	Value per share	Dividend per share
AKT Limited	1200	$123.56	$8.45
Takit Group	7000	$56.35	$3.81
Metals Ltd	54000	$2.54	$0.14
Jeluvion Fund	16500	$8.45	$0.28

3 Portfolio Three

Company	Number of shares	Value per share	Dividend per share
Tekan Limited	5 000	$4.78	$0.23
TPSD Inc	2 300	$7.92	No dividend
Superla Fund	11 800	$1.23	$0.08
BKJ Resources	3 500	$4.80	$0.45
Lumsdon Corp	450	$67.55	$3.24

4 Portfolio Four

Company	Number of shares	Value per share	Dividend per share
Calivia Ltd	2000	$5.60	8.2% of share price
Mavis and Co	2000	$3.85	5.3% of share price
Dally Inc	500	$17.80	3.8% of share price
Tuscan Corp	1200	$9.56	10.2% of share price
Petron Oil Ltd	4000	$2.67	No dividend
Parack Trust	6500	$0.85	6.3% of share price
Savings Trust	3200	$6.85	3.6% of share price
Borek Limited	1500	$23.40	2.7% of share price

5 Determine the P/E (the price-to-earnings ratio) of a company that has a share price of $54 and for which the total earnings over the previous twelve months has been $4.50.

6 Determine the P/E (the price-to-earnings ratio) of a company that has a share price of $12.80 and for which the total earnings over the previous twelve months has been $0.95.

7 List the following companies in order from the one with the lowest P/E (price-to-earnings ratio) first to the one with the highest P/E last.

Company	Current share price	Total dividend in last 12 months
DeepGas Limited	$54.45	$4.36
Iron Resources	$16.86	$1.22
Jupiter Trust	$0.92	$0.08
Linear Corp	$85.40	$3.85
Japatali Fund	$7.84	$0.82
Premiere Bank	$18.56	$2.47
Tacomala Group	$3.48	$0.28

8 a A company declares its annual dividend equal to 10% of its share price. What is the company's price to earnings ratio?

b A company declares its annual dividend equal to 5% of its share price. What is the company's price to earnings ratio?

c If a company has a price to earnings ratio of 8 express its current dividend per year as a percentage of the share price.

9 The price of a share in a company can fall a little when it goes 'ex-dividend'. Explain (research if necessary).

Government allowances and pensions

Raising children can be expensive. Some parents on low incomes can find it difficult to pay all the bills from their income. In such cases the Government *Family Tax Benefit* may help.

When a person reaches an age at which they retire from work the money they were earning from working stops but many of the bills they incur, food bills for example, still need to be paid. In such cases the Government's *Age Pension* may help.

If you need to provide care to a person in your family who has a physical, intellectual or psychiatric disability it could be difficult for you to also have a full time job. Without this ability to work full time you may have trouble performing your duties as a carer and paying the bills and expenses. In such cases the Government's *Carer Payment* may help, or perhaps the *Carer Allowance* may be of assistance.

If someone aged 25 or over wants to enrol for a full time education or training course they could find it very difficult to pay the everyday bills when their training commitment does not leave much time for a paid job. In such cases the Government *Austudy* may help.

If someone has a physical, intellectual or psychiatric impairment that limits their ability to undertake work, or training for work, then their ability to earn sufficient money to meet their living expenses could also be limited. In such cases the Government's *Disability Support Pension* may help.

Payments like the Family Tax Benefit, the Age Pension, Carer Payments, Carer Allowance, Austudy and the Disability Support Pension mentioned above, and others, are available to financially help those in need. However the Government has to ensure that the people who get these payments are genuinely in need of them. To be eligible for such payments people have to meet basic conditions, such as being an Australian resident, being of a certain age, being enrolled for particular courses, having particular disabilities etc. In addition the payments are often subject to an income test, and in some cases an assets test. If a person's income is above a particular level the allowance may be reduced or perhaps denied altogether. Similar reductions may occur if the total value of the assets a person owns exceeds particular amounts.

To get some familiarity with these ideas this section will consider three Government allowances: Carer Allowance, Family Tax Benefit and The Age Pension.

ISBN 9780170390194

Note:
- When considering these allowances this text may present a somewhat simplified version of the real system in order to make calculations more manageable, whilst still retaining the basic principles of the process.
- The allowances and rules shown here should *not* be taken as the situation that necessarily applies at the time you are working through this topic.

Carer Allowance

If a person provides care to an adult or dependent child who's disability, medical condition or frailty means that they require care on a daily basis then the person may be eligible to receive a *Carer Allowance*, provided both care giver and care receiver are Australian residents. (If the caring required is more constant the carer may qualify for the more substantial Carer Payment, not considered here.)

The Carer Allowance is a payment of $115.40 per fortnight plus a once a year lump sum payment of $600 paid every July.

The carer can receive this Care Allowance irrespective of the income or assets of the care giver or of the care receiver.

Family Tax Benefit

To help with the cost of bringing up children a Family Tax Benefit can be paid to the parents or guardians of any child who is up to 15 years of age, or is between 15 and 19 and still in full time secondary study, and who is living with the parent or guardian making the claim for benefit. How much will be paid depends on the total annual income of the parents or guardians and the number of children in the family who meet the criteria.

(In the real system, the allowance depends on the age of the child but, to keep it straightforward, this text will not make this distinction and will also only consider the situation for families with one or two children.)

Families with 1 child meeting the criteria	
Combined annual income	**Family Tax Benefit for the year**
Up to $48 000	$5100
$48 000 to $63 000	$5100 less 20 cents for each $1 annual income exceeds $48 000
$63 000 to $95 000	$2100
$95 000 to $102 000	$2100 less 30 cents for each $1 that annual income exceeds $95 000
Over $102 000	Nil

Families with 2 children meeting the criteria	
Combined annual income	**Family Tax Benefit for the year**
Up to $48 000	$10 200
$48 000 to $78 000	$10 200 less 20 cents for each $1 annual income exceeds $48 000
$78 000 to $98 000	$4200
$98 000 to $112 000	$4200 less 30 cents for each $1 that annual income exceeds $98 000
Over $112 000	Nil

The Age Pension

Once a person reaches retirement age, which for someone born after 1st January 1957 is 67, they may be entitled to receive the *Age Pension*. Whether the aged person receives all or any of the pension depends on their total income from all other sources and the total value of their assets and those of their partner, excluding the value of their house if they own one. They must also have been an Australian resident for a total of at least ten years, of which at least five must have been unbroken.

For the purposes of this text the fortnightly pension rates should be taken to be as given below. However these are used to illustrate the process only. For actual rates, amounts and rules the reader should consult the appropriate government websites.

$772.60 per fortnight for a single person.
(Consisting of a base amount of $712 + a pensioner supplement of $60.60)

$582.40 per fortnight for each qualifying member of a couple.
(Consisting of a base amount of $536.70 + a pensioner supplement of $45.70)

However these amounts may be reduced subject to the income test and the assets test. Both of these tests are applied to assess the level of payment and the one that gives the lower level of payment will be the one that applies.

The Assets test

For singles and couples, who are or are not homeowners, the assets limit for receiving the full pension should be taken to be as follows:

Situation	Homeowners	Non-homeowners
Single	$192 500	$332 000
Couple (combined assets)	$273 000	$412 500

Assets above the amounts shown in the table will reduce the pension by $1.50 per fortnight ($0.75 for each member of a couple) for each $1000 above the amount shown in the table.

Note: Any of the assets that are income earning assets, e.g. bank accounts, shares etc, will be deemed to earn income at the 'deeming rate' – we will use a 4% per annum deeming rate.

The Income test

For singles and couples the income limit for receiving a full pension should be taken as:

Situation	
Single	$152.00 per fortnight
Couple (combined income)	$268.00 per fortnight

Fortnightly income above the amounts shown in the table will reduce the pension by 50 cents per fortnight (25 cents for each member of a couple) for each $1 above that shown in the table. However of the income that is earned from employment (rather than deemed income from assets like bank accounts etc.) the first $250 earned per fortnight does not count in the income test.

ISBN 9780170390194

EXAMPLE 4

Determine the fortnightly age pension for a single homeowner of pension age with assets of $257 000 (excluding the family home), of which $150 000 will be deemed to earn 4% per annum income, and a fortnightly earned income of $680.

Solution

Under the Assets test:

$$\text{Fortnightly pension} = \$772.60 - \frac{\$257\,000 - \$192\,500}{\$1000} \times \$1.50$$
$$= \$675.85$$

Under the Income test:

Earned income of $680 per fortnight + deemed income of $\dfrac{\$150\,000 \times 0.04}{26}$ per fortnight.

Thus for the Income test we count fortnightly income of ($680 – $250) + $231 (rounded)

$$= \$661$$

$$\text{Fortnightly pension} = \$772.60 - (\$661 - \$152) \times \$0.50$$
$$= \$518.10$$

Choosing the lesser of these gives the fortnightly pension as $518 (nearest dollar).

EXAMPLE 5

Determine the fortnightly age pension for each pension age member of a homeowning couple with assets of $750 000 (excluding the family home), of which $650 000 will be deemed to earn 4% per annum income, and a fortnightly earned income of $120.

Solution

Under the Assets test:

$$\text{Fortnightly pension} = \$582.40 - \frac{\$750\,000 - \$273\,000}{\$1000} \times \$0.75$$
$$= \$224.65$$

Under the Income test:

Earned income of $120 per fortnight + deemed income of $\dfrac{\$650\,000 \times 0.04}{26}$ per fortnight.

Thus for the Income test we count fortnightly income of ($120 – $120) + $1000

$$= \$1000$$

$$\text{Fortnightly pension} = \$582.40 - (\$1000 - \$268) \times \$0.25$$
$$= \$399.40$$

Choosing the lesser of these gives the fortnightly pension as $225 (nearest dollar) for each member of the couple meeting pension age criteria.

Exercise 5E

For this exercise you will need to refer to the various allowance details given on pages 75–76. Any assets stated for homeowners should be assumed to exclude the family home.

1 What is the annual Carer Allowance?

2 Determine the annual Family Tax Benefit (FTB) paid to a family with one child meeting the FTB requirements and with a combined annual family income of $52 000.

3 Determine the fortnightly Family Tax Benefit (FTB), paid to a family with two children meeting the FTB requirements and with a combined annual family income of $135 000.

4 Determine the fortnightly Family Tax Benefit (FTB) paid to a family with two children meeting the FTB requirements and with a combined annual family income of $102 000.

5 Determine the annual Family Tax Benefit (FTB) paid to a family with one child meeting the FTB requirements and with a combined annual family income of $37 000.

6 Determine the fortnightly age pension for a single homeowner of pension age with assets of $650 000, of which $450 000 will be deemed to earn 4% per annum income, and a fortnightly earned income of $120.

7 Determine the fortnightly age pension for each pension age member of a homeowning couple with assets of $220 000, of which $185 000 will be deemed to earn 4% per annum income, and a fortnightly earned income of $300.

8 Determine the fortnightly age pension for each pension age member of a non-homeowning couple with assets of $320 000, of which $250 000 will be deemed to earn 4% per annum income, and an annual earned income of $22 100.

9 Determine the fortnightly age pension for a single non-homeowner of pension age with assets of $90 000, of which $70 000 will be deemed to earn 4% per annum income, and a fortnightly earned income of $400.

10 Determine the fortnightly age pension for each pension age member of a non-homeowning couple with assets of $245 000, of which $150 000 will be deemed to earn 4% per annum income, and an annual earned income of $9074.

11 Determine the fortnightly age pension for a single homeowner of pension age with assets of $550 000, of which $500 000 will be deemed to earn 4% per annum income, and a fortnightly earned income of $250.

12 Determine the fortnightly age pension for a single non-homeowner of pension age with assets of $25 000, of which $5000 will be deemed to earn 4% per annum income, and a fortnightly earned income of $800.

MATHEMATICS APPLICATIONS Unit 1 ISBN 9780170390194

Budgeting

Pete and Julie want to take themselves and their two children away on a holiday next year but are not sure they will be able to save enough money to pay for it. They decide to prepare a budget. This means that they allocate future earnings to the various household expenditures likely to occur, in this case to see if this planned expenditure leaves enough for a holiday fund to be created. They know that the total family income is $1330 per week and they list the following likely weekly expenditure items:

Budget grid

Budgeting scenarios

Rent $320	Gas and electricity $30	Phones $20
Internet $15	Entertainment/Eating out $100	Clothing $75
Car $160	Bus/train travel $20	Insurances $40
Loan repayments $95	Food $200	Sports and Gym $50
Health fund $80		

Some of the above items involve **fixed expenditure**. For example the internet costs or the rent, which would be set at a fixed, agreed amount per week (probably paid on a monthly basis).

Some of the items involve **discretionary spending**. For example the allowance for entertainment/ eating out is something the family have some *discretion* over, i.e. the family can choose to spend more or less than this amount. It is not a fixed amount.

The budget shown below indicates that Pete and Julie could start a holiday fund with $125 being available to go into it each week:

Weekly budget			
Income		**Expenditure**	
Family benefit	$80	Bus/train travel	$20
Wages & Salary (after tax)	$1250	Car	$160
		Clothing	$75
		Ent./Eating out	$100
		Food	$200
		Gas & Elec	$30
		Health fund	$80
		Internet	$15
		Insurances	$40
		Loan repayment	$95
		Phones	$20
		Rent	$320
		Sports & Gym	$50
Total Income	**$1330**	**Total Expenditure**	**$1205**
		Income – Expenditure	**$125**

TECHNOLOGY

A spreadsheet can be particularly useful in creating a budget.

Note: A household budget gives an expenditure guideline and expenditure on each item should be kept within those guidelines if possible. However unexpected expenditure can be necessary so some flexibility should be allowed for. If the budget overspends a little on one item, savings can be looked for in another area, or savings can be looked for from the budget of the following week, or the budget may need to be adjusted, etc.

Over time the budget will need to be adjusted to reflect changes to income, expenditure patterns, family circumstances etc.

Exercise 5F

1 Discuss with others in your class how realistic the figures in the previous budget seem. Do any of the amounts shown strike you as being surprisingly high/low? Any areas where you think money could be saved or further money would need to be spent? Any items of likely expenditure missing?

2 Imagine you are a parent in a family comprising Mum, Dad and two children (and a cat).

The total income for the family is $128 000 of which 80% is available as 'take home pay' after income tax has been deducted. The total salary is due to one of the parents working full time and the other working part time. The two children are both of school age but some before school and after school child care facilities have to be allowed for.

Think of the likely expenses such a family will incur and, with the assistance of a spreadsheet, prepare a fortnightly budget for the family.

3 Think of your own income and expenditure (or that of a fictitious friend of about your age who attends school and either does some part time evening and weekend work or receives a financial allowance from parents). Prepare a weekly budget for yourself (or the fictitious friend).

4 There are a number of internet budget planners available. Investigate.

5 Suzanne is 20 years old and thinking of purchasing a car. She has enough saved to be able to do this but as part of preparing her budget she wants to know what the likely running costs are e.g. fuel costs, insurance, road tax, servicing etc.

Investigate and prepare a written report for her. Include a spreadsheet display in your report.

Miscellaneous exercise five

This miscellaneous exercise may include questions involving the work of this chapter, the work of any previous chapters, and the ideas mentioned in the Preliminary work at the beginning of the book.

1 Which is the greater in each of the following:

   a 80% of $55 or 9.6% of $450.   **b** 126% of $70 or 93% of $95.

   c 80% of $50 or 50% of $80.   **d** 40% of $65 or 65% of $40.

2 Determine the value after three years of an investment of $5000 invested at 6% per annum compound interest with the compounding occurring

   a annually   **b** every six months   **c** quarterly   **d** monthly.

3 With 'all other things being equal' the cost of each of the following blocks of land could be compared and ranked in order of 'best value for money' on the basis of cost per square metre. Rank the blocks in this way.

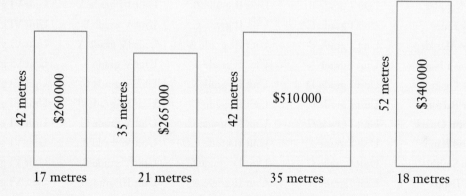

In this situation how likely is it that 'all other things will be equal'.

4 If you are 22 years of age, or older, and are unemployed but actively seeking paid work, and meet certain income and assets tests, you may be eligible for the Government *Newstart Allowance*.

If you are single, with no children, and earn less than $62 per fortnight, *Newstart* gives you an allowance of $492.60 per fortnight.

This fortnightly allowance reduces by 50 cents in the dollar for each dollar you earn over $62 per fortnight, up to you earning $250 per fortnight. Income above $250 per fortnight further reduces the allowance by 60 cents in the dollar for each dollar you earn over $250 per fortnight.

   a Calculate the fortnightly Newstart Allowance for a single 24 year old, with no children, who meets the criteria for receiving the Newstart Allowance and earns $100 per fortnight.

   b Calculate the fortnightly Newstart Allowance for a single 28 year old, with no children, who meets the criteria for receiving the Newstart Allowance and earns $300 per fortnight.

   c As the fortnightly income increases (in whole dollars) at what amount does the Newstart Allowance first 'cut out' altogether for a single person, over 22, with no children?

5 The 25 Engineering students at a college all do 4 units chosen from 6.

All the students have to do unit I but then choose 3 of II, III, IV, V and VI.

They are then awarded grades A, B, C, D or F in each unit.

The results for the 25 students are shown below.

Alan Amos	Unit I grade B	Unit IV grade A	Unit V grade B	Unit VI grade B
Betty Baxter	Unit I grade B	Unit II grade C	Unit III grade B	Unit IV grade B
Carlie Crabb	Unit I grade C	Unit III grade B	Unit IV grade C	Unit VI grade B
Diane Denny	Unit I grade B	Unit IV grade B	Unit V grade B	Unit VI grade B
Eric Even	Unit I grade C	Unit II grade C	Unit V grade B	Unit VI grade B
Frank Fermat	Unit I grade C	Unit II grade D	Unit IV grade D	Unit V grade C
Gwen Garland	Unit I grade C	Unit II grade D	Unit V grade D	Unit VI grade C
Harry Hughes	Unit I grade F	Unit III grade F	Unit V grade F	Unit VI grade D
Ian Icon	Unit I grade B	Unit II grade C	Unit IV grade B	Unit VI grade B
Janice Jones	Unit I grade D	Unit III grade C	Unit IV grade D	Unit V grade C
Kim Keppler	Unit I grade B	Unit II grade B	Unit III grade A	Unit VI grade A
Larry Lines	Unit I grade C	Unit II grade C	Unit V grade B	Unit VI grade C
Mavis Marsden	Unit I grade B	Unit III grade A	Unit IV grade A	Unit VI grade B
Norman Napp	Unit I grade A	Unit II grade A	Unit V grade A	Unit VI grade A
Olive Ogate	Unit I grade D	Unit IV grade C	Unit V grade C	Unit VI grade C
Peter Parton	Unit I grade C	Unit II grade C	Unit IV grade B	Unit VI grade D
Quintin Quark	Unit I grade C	Unit IV grade C	Unit V grade C	Unit VI grade C
Roland Rapp	Unit I grade A	Unit II grade C	Unit V grade B	Unit VI grade B
Sadie Smith	Unit I grade C	Unit IV grade C	Unit V grade B	Unit VI grade B
Tom Tripp	Unit I grade A	Unit II grade A	Unit III grade B	Unit VI grade A
Una Userp	Unit I grade C	Unit III grade D	Unit V grade D	Unit VI grade D
Victor Velor	Unit I grade D	Unit II grade D	Unit V grade D	Unit VI grade D
Walter Wong	Unit I grade C	Unit III grade C	Unit IV grade B	Unit VI grade C
Xue Xenides	Unit I grade A	Unit II grade B	Unit III grade A	Unit VI grade A
Ying Yenson	Unit I grade B	Unit III grade B	Unit IV grade B	Unit VI grade A

As part of the process of comparing the units, the above data is used to complete the table shown below. Copy and complete this table, giving percentages to the nearest integer.

	Number of students	Percentage of those doing the unit getting				
		As	Bs	Cs	Ds	Fs
Unit I						
Unit II						
Unit III						
Unit IV						
Unit V						
Unit VI						

ISBN 9780170390194

Matrices

Situation

A league soccer competition involves six teams:

Ajax

Battlers

Cloggers

Devils

Enzymes

Flames

pixabay.com/jarmoluk

Each team plays one game per week and, during the ten week competition, plays each other team twice, once in the first five weeks and once in the last five weeks. (All teams play at the same venue so no consideration needs to be made to balance home and away games.)

The results for the first five weeks gave rise to the following table:

	Played	Won	Drawn	Lost	Goals scored	
					For	Against
Ajax	5	2	1	2	10	5
Battlers	5	2	1	2	4	5
Cloggers	5	2	0	3	7	6
Devils	5	2	0	3	4	11
Enzymes	5	3	2	0	8	2
Flames	5	2	0	3	5	9

- Create a similar table for the last five weeks of the competition using the results stated below.

Week 6

Ajax 3 1 Battlers

Cloggers 4 1 Devils

Enzymes 5 4 Flames

Week 7

Ajax 1 2 Cloggers

Battlers 1 0 Enzymes

Devils 1 1 Flames

Week 8

Ajax 2 2 Devils

Battlers 1 0 Flames

Cloggers 4 3 Enzymes

Week 9

Ajax 0 1 Enzymes

Battlers 2 0 Devils

Cloggers 1 0 Flames

Week 10

Ajax 1 3 Flames

Battlers 0 1 Cloggers

Devils 0 1 Enzymes

- Create a table like the one above for the complete ten-week competition.

As part of the soccer league activity on the previous page we had to arrange information in a 'rows and columns' form of presentation.

This rows and columns *rectangular array* presentation of numbers is called a **matrix**. (Plural: matrices).

If we remove the headings and indicate the start and end of the matrix with brackets, the table given on the previous page would be written as shown on the right.

$$\begin{bmatrix} 5 & 2 & 1 & 2 & 10 & 5 \\ 5 & 2 & 1 & 2 & 4 & 5 \\ 5 & 2 & 0 & 3 & 7 & 6 \\ 5 & 2 & 0 & 3 & 4 & 11 \\ 5 & 3 & 2 & 0 & 8 & 2 \\ 5 & 2 & 0 & 3 & 5 & 9 \end{bmatrix}$$

This matrix has 6 rows and 6 columns. We say it is a *six by six* matrix, (written 6×6). This gives the **size** or **dimensions** of the matrix.

Matrices do not have to have the same number of rows as they have columns, however those that do are called **square matrices**.

The matrix on the right is a 6×6 square matrix.

$$\begin{bmatrix} 1 & 0 & 4 \\ 3 & 2 & 0.5 \end{bmatrix} \qquad \begin{bmatrix} 2 & 5 \\ 11 & -2 \end{bmatrix} \qquad \begin{bmatrix} 3 \\ 2 \\ -1 \end{bmatrix} \qquad \begin{bmatrix} 1 & 0 & 2 & 3 \\ 1 & 2 & -1 & 0 \\ 0 & 1 & 0 & 3 \end{bmatrix}$$

2 rows and 3 columns.   2 rows and 2 columns   3 rows and 1 column.   3 rows and 4 columns.
A 2×3 matrix.   A 2×2 matrix.   A 3×1 matrix.   A 3×4 matrix.
                      (A square matrix.)

A matrix consisting of just one column, like the third matrix above, is called a **column matrix**.

Any matrix consisting of just one row is called a **row matrix**. For example:

$$\begin{bmatrix} 5 & 0 & -2 & 1 \end{bmatrix}$$

A square matrix having zeros in all spaces that are not on the **leading diagonal** is called a **diagonal matrix**.

$$\begin{bmatrix} 5 & 0 & 0 & 0 \\ 0 & 0 & 0 & 0 \\ 0 & 0 & 2 & 0 \\ 0 & 0 & 0 & -4 \end{bmatrix}$$

We commonly use capital letters to label different matrices. The corresponding lower case letters, with subscripted numbers, are then used to indicate the row and column a particular entry or **element** occupies.

For the matrix A shown on the right the element occupying the 3rd row and 2nd column is the number 7.

$$A = \begin{bmatrix} 2 & 0 & -1 & 3 \\ 1 & 6 & -3 & 9 \\ 5 & 7 & 8 & 4 \end{bmatrix}$$

Thus $a_{32} = 7$.

Similarly $a_{11} = 2$,

                 $a_{12} = 0$,

                 $a_{13} = -1$, etc.

Adding and subtracting matrices

In the soccer competition activity earlier in this chapter you probably determined the matrix for the full ten weeks by adding the matrix for the first five weeks to the matrix for the last five weeks. To perform such addition it was natural to simply add elements occurring in corresponding locations. This is indeed how we add matrices. For example,

$$\text{if } A = \begin{bmatrix} 1 & 0 & 2 & 3 \\ 4 & -2 & 3 & 5 \\ 2 & 1 & -3 & 4 \end{bmatrix} \quad \text{and } B = \begin{bmatrix} 2 & 1 & -3 & 2 \\ 5 & 1 & 2 & 4 \\ 3 & 2 & 0 & -5 \end{bmatrix} \quad \text{then } A + B = \begin{bmatrix} 3 & 1 & -1 & 5 \\ 9 & -1 & 5 & 9 \\ 5 & 3 & -3 & -1 \end{bmatrix}$$

$$\text{and similarly} \qquad A - B = \begin{bmatrix} -1 & -1 & 5 & 1 \\ -1 & -3 & 1 & 1 \\ -1 & -1 & -3 & 9 \end{bmatrix}$$

Note: When adding or subtracting matrices there must be elements in corresponding spaces. Thus we can only add or subtract matrices that are the same size as each other.

Multiplying a matrix by a number

Suppose that the 3×2 matrix shown on the right shows the cost of three models of gas heater in two different shops.

Now suppose that in a sale both shops offer 10% discount on all models.

	Shop One	Shop Two
Economy	$250	$280
Standard	$340	$330
Deluxe	$450	$450

The sale prices could be represented in a matrix formed by multiplying each element of the first matrix by 0.9.

	Shop One	Shop Two
Economy	$225	$252
Standard	$306	$297
Deluxe	$405	$405

This is indeed how we multiply a matrix by a number: we multiply each element of the matrix by that number. (This is referred to as 'multiplication by a **scalar**'.)

Shutterstock.com/Kenneth William Caleno

Equal matrices

For two matrices to be equal they must be of the same size and have all corresponding elements equal.

Thus if $\begin{bmatrix} a & b & c \\ d & e & f \end{bmatrix} = \begin{bmatrix} 2 & 3 & -5 \\ 1 & 0 & -2 \end{bmatrix}$ then $\begin{array}{ccc} a=2 & b=3 & c=-5 \\ d=1 & e=0 & f=-2 \end{array}$

EXAMPLE 1

If $A = \begin{bmatrix} 1 & 2 & 4 \\ 0 & -4 & 5 \end{bmatrix}$, $B = \begin{bmatrix} 3 & 5 & -2 \\ 1 & 0 & -2 \end{bmatrix}$ and $C = \begin{bmatrix} 2 & 3 \\ 1 & -5 \end{bmatrix}$ determine each of the following.

If any cannot be determined, state this clearly and give the reason.

a $A + B$ **b** $A + C$ **c** $B - A$ **d** $5C$ **e** $3B - 2A$

Solution

a $A + B = \begin{bmatrix} 1 & 2 & 4 \\ 0 & -4 & 5 \end{bmatrix} + \begin{bmatrix} 3 & 5 & -2 \\ 1 & 0 & -2 \end{bmatrix}$

$= \begin{bmatrix} 4 & 7 & 2 \\ 1 & -4 & 3 \end{bmatrix}$

b A and C are not the same size. Thus A + C cannot be determined.

c $B - A = \begin{bmatrix} 3 & 5 & -2 \\ 1 & 0 & -2 \end{bmatrix} - \begin{bmatrix} 1 & 2 & 4 \\ 0 & -4 & 5 \end{bmatrix}$ **d** $5C = \begin{bmatrix} 10 & 15 \\ 5 & -25 \end{bmatrix}$

$= \begin{bmatrix} 2 & 3 & -6 \\ 1 & 4 & -7 \end{bmatrix}$

e $3B - 2A = \begin{bmatrix} 9 & 15 & -6 \\ 3 & 0 & -6 \end{bmatrix} - \begin{bmatrix} 2 & 4 & 8 \\ 0 & -8 & 10 \end{bmatrix}$

$= \begin{bmatrix} 7 & 11 & -14 \\ 3 & 8 & -16 \end{bmatrix}$

Many calculators will accept data in matrix form and can then manipulate these matrices in various ways.

Get to know the matrix capability of your calculator.

How does your calculator respond when you ask it to add together two matrices that are not of the same size?

$\begin{bmatrix} 1 & 2 & 4 \\ 0 & -4 & 5 \end{bmatrix} + \begin{bmatrix} 3 & 5 & -2 \\ 1 & 0 & -2 \end{bmatrix}$

$\begin{bmatrix} 4 & 7 & 2 \\ 1 & -4 & 3 \end{bmatrix}$

Exercise 6A

1 A matrix with m rows and n columns has size m × n. Write down the size of each of the following matrices.

$$A = \begin{bmatrix} 2 & -3 \\ 3 & 4 \\ 1 & 0 \\ 2 & -1 \end{bmatrix} \qquad B = \begin{bmatrix} 2 & 3 & -7 & 32 \\ 1 & 1 & 0 & 5 \end{bmatrix} \qquad C = \begin{bmatrix} -1 \\ 3 \\ 5 \\ 2 \end{bmatrix} \qquad D = \begin{bmatrix} 1 & 0 & 0 \\ 3 & 0 & 1 \\ 0 & 5 & 0 \\ 0 & 1 & 3 \end{bmatrix}$$

$$E = \begin{bmatrix} 2 & 5 \\ 0 & 1 \end{bmatrix} \qquad F = \begin{bmatrix} 1 & 11 & -2 \end{bmatrix} \qquad G = \begin{bmatrix} 12 & 3 \\ 0 & 5 \\ -5 & 2 \end{bmatrix} \qquad H = \begin{bmatrix} 1 & 0 & 0 & 0 \\ 0 & 1 & 0 & 0 \\ 0 & 0 & 1 & 0 \\ 0 & 0 & 0 & 1 \end{bmatrix}$$

2 If e_{mn} is the element situated in the mth row and nth column of matrix E determine

a e_{12} **b** e_{21} **c** f_{13}

d g_{21} **e** g_{22} **f** g_{32}

where matrices E, F and G are as given below.

$$E = \begin{bmatrix} 5 & 4 & 13 \\ -4 & 2 & 0 \\ 1 & -8 & 12 \end{bmatrix} \qquad F = \begin{bmatrix} 1 & 5 & 7 & 2 \end{bmatrix} \qquad G = \begin{bmatrix} 1 & 2 \\ 7 & 3 \\ -2 & 0 \\ 4 & 11 \end{bmatrix}$$

3 If $A = \begin{bmatrix} 1 & 2 \\ 0 & -4 \end{bmatrix}$, $B = \begin{bmatrix} 3 & -1 \\ 2 & 4 \\ 0 & 3 \end{bmatrix}$, $C = \begin{bmatrix} 2 & -3 \\ 1 & -5 \end{bmatrix}$ and $D = \begin{bmatrix} 3 \\ 1 \\ -2 \end{bmatrix}$, determine each of the

following. If any cannot be determined state this clearly.

a A + B   **b** A + C   **c** C − A   **d** 2D

e 3B   **f** B + D   **g** 2A   **h** 2A − C

4 If $P = \begin{bmatrix} 3 & 2 & -1 \\ 1 & 4 & 3 \end{bmatrix}$, $Q = \begin{bmatrix} 2 & 1 & 0 \\ 0 & -1 & 0 \end{bmatrix}$ and $R = \begin{bmatrix} 1 & 2 & 1 \\ 2 & 1 & 2 \end{bmatrix}$ determine each of the following.

If any cannot be determined state this clearly.

a P + Q   **b** Q − P   **c** 3R   **d** 3P − 2Q

5 If $A = \begin{bmatrix} 2 & 4 \\ 1 & 3 \end{bmatrix}$, $B = \begin{bmatrix} 2 & 1 & 3 \end{bmatrix}$, $C = \begin{bmatrix} 3 & 1 & 4 \end{bmatrix}$ and $D = \begin{bmatrix} 2 \\ 1 \\ 3 \end{bmatrix}$ determine each of the following.

If any cannot be determined state this clearly.

a A + B   **b** 3A   **c** B + 2C   **d** C + D

6 If $A = \begin{bmatrix} 1 & 3 & 0 & 1 \\ 0 & 1 & 2 & 3 \\ 0 & 0 & 1 & 4 \end{bmatrix}$, $B = \begin{bmatrix} 3 & 1 & 4 \\ 2 & 1 & -3 \\ 0 & 1 & 2 \\ 1 & 0 & 0 \end{bmatrix}$ and $C = \begin{bmatrix} 5 & 1 & 3 & -1 \\ 2 & 1 & 4 & 3 \\ 1 & 5 & 2 & 0 \end{bmatrix}$ determine each of the

following. If any cannot be determined state this clearly.

   a $A + B$ **b** $A + C$ **c** $2B$ **d** $5A - C$

7 $A = \begin{bmatrix} 1 & 2 \end{bmatrix}$, $B = \begin{bmatrix} 3 \\ 1 \end{bmatrix}$, $C = \begin{bmatrix} 5 & 2 \end{bmatrix}$ and $D = \begin{bmatrix} 1 & 7 \end{bmatrix}$.

For each of the following write 'Yes' if it can be determined and 'No' if it cannot be determined.

   a $A + B$ **b** $B - A$ **c** $3C$ **d** $A + D$

   e $A - 3D$ **f** $A + 3B$ **g** $B + B$ **h** $A + B + C$

8 Is matrix addition commutative? i.e. Does $A + B = B + A$ (assuming A and B are of the same size)?

9 Is matrix addition associative? i.e. Does $A + (B + C) = (A + B) + C$ (assuming A, B and C are all of the same size)?

10 If $A = \begin{bmatrix} 1 & -1 & 2 \\ 1 & 0 & 3 \end{bmatrix}$ and $B = \begin{bmatrix} 1 & -7 & 12 \\ 1 & 0 & 13 \end{bmatrix}$ determine matrix C given that the following equation

is correct:

$$3A - 2C = B.$$

11 For the first four games in a basketball season the points (P), assists (A), and blocks (B), that five members of one team carried out were as shown below.

Game 1

	P	A	B
Alan	8	5	1
Bob	7	2	4
Dave	14	3	1
Mark	17	3	1
Roger	8	8	2

Game 2

	P	A	B
Alan	12	3	1
Bob	6	8	2
Dave	15	3	5
Mark	6	4	0
Roger	5	2	6

Game 3

	P	A	B
Alan	11	8	2
Bob	15	2	5
Dave	7	5	2
Mark	12	2	1
Roger	14	4	5

Game 4

	P	A	B
Alan	9	4	0
Bob	9	3	3
Dave	11	8	1
Mark	4	12	1
Roger	12	5	3

   a   Construct a single 5×3 matrix showing the total points, total assists and total blocks each of these five players achieved for the 4 game period.

   b   Construct a single 5×3 matrix showing the average points per game, average assists per game and average blocks per game for each of these five players for the 4 game period.

12 A company manufactures five types of lawn fertiliser:

- Basic (B)   • Feedit (F)   • Fertilawn (FL)   • Greenit (G)   • Growgrass (GG)

It sells these through its four garden centres.

The number of bags of these fertilisers sold in these centres during the first and second halves of a year are given below:

$$
\begin{array}{c}
\text{1 January} \rightarrow \text{30 June} \\
\begin{array}{l}
\quad\quad\quad\text{B}\quad\text{F}\quad\text{FL}\quad\text{G}\quad\text{GG} \\
\begin{array}{l}
\text{Centre I} \\
\text{Centre II} \\
\text{Centre III} \\
\text{Centre IV}
\end{array}
\left[
\begin{array}{ccccc}
3100 & 550 & 1040 & 820 & 2250 \\
1640 & 420 & 720 & 480 & 1480 \\
2850 & 520 & 1320 & 640 & 1250 \\
1240 & 300 & 800 & 360 & 960
\end{array}
\right]
\end{array}
\end{array}
$$

$$
\begin{array}{c}
\text{1 July} \rightarrow \text{31 December} \\
\begin{array}{l}
\quad\quad\quad\text{B}\quad\text{F}\quad\text{FL}\quad\text{G}\quad\text{GG} \\
\begin{array}{l}
\text{Centre I} \\
\text{Centre II} \\
\text{Centre III} \\
\text{Centre IV}
\end{array}
\left[
\begin{array}{ccccc}
2500 & 1200 & 1280 & 950 & 2000 \\
1200 & 850 & 650 & 540 & 1240 \\
2200 & 950 & 1500 & 640 & 1450 \\
950 & 640 & 720 & 480 & 820
\end{array}
\right]
\end{array}
\end{array}
$$

The company predicts that at each shop the sales for the next year will increase by 10% due to a new sales campaign. Assuming this prediction is indeed correct produce a 4×5 matrix showing the number of bags of each fertiliser sold at each shop for the following 1 January $\rightarrow$ 31 December. (Use a spreadsheet or the ability of some calculators to manipulate matrices if you wish.)

13 If a_{mn} is the element situated in the mth row and nth column of matrix A write down matrix A given that it is a 3×3 matrix with $a_{mn} = 2m + n$.

14 If a_{mn} is the element situated in the mth row and nth column of matrix A write down matrix A given that it is a 3×4 matrix with $a_{mn} = m^n$.

Multiplying matrices

An inter-school sports carnival involves five schools competing in seven sports.

In each of these sports, medals, certificates and team points are awarded to teams finishing 1st, 2nd or 3rd.

The 5×3 matrix on the right shows the number of first, second and third places gained by each of the five schools.

$$
\begin{array}{l}
\quad\quad\quad\quad\text{1st Place}\quad\text{2nd Place}\quad\text{3rd Place} \\
\begin{array}{l}
\text{School A} \\
\text{School B} \\
\text{School C} \\
\text{School D} \\
\text{School E}
\end{array}
\left[
\begin{array}{ccc}
1 & 1 & 1 \\
3 & 1 & 0 \\
0 & 3 & 3 \\
1 & 2 & 0 \\
2 & 0 & 3
\end{array}
\right]
\end{array}
$$

Suppose that points are awarded using the points system:

The total points scored for each school are:

$$
\begin{array}{l}
\text{1st} \\
\text{2nd} \\
\text{3rd}
\end{array}
\left[
\begin{array}{l}
\text{3 points} \\
\text{2 points} \\
\text{1 point}
\end{array}
\right]
$$

School A	School B	School C	School D	School E
$1 \times 3 +$	$3 \times 3 +$	$0 \times 3 +$	$1 \times 3 +$	$2 \times 3 +$
$1 \times 2 +$	$1 \times 2 +$	$3 \times 2 +$	$2 \times 2 +$	$0 \times 2 +$
1×1	0×1	3×1	0×1	3×1
6	11	9	7	9

Thus school B finished first with 11 points, followed by schools C and E equal second, school D was fourth and school A was fifth.

Note the way that each row of the 5×3 matrix is 'stood up' to align with the points matrix. This is indeed how we carry out matrix multiplication.

Matrices can be multiplied together if the number of columns in the first matrix equals the number of rows in the second matrix.

If $A = \begin{bmatrix} 2 & 1 & 3 \\ 0 & -1 & 2 \end{bmatrix}$ and $B = \begin{bmatrix} 1 & 2 \\ 4 & -1 \\ 1 & -3 \end{bmatrix}$ the product AB is found as shown below.

Follow each step carefully to make sure you understand where each element in the final answer comes from.

First spin the 1st row of A to align with 1st column of B, multiply and then add:

Thus $\begin{bmatrix} \mathbf{2} & \mathbf{1} & \mathbf{3} \\ 0 & -1 & 2 \end{bmatrix} \begin{bmatrix} \mathbf{1} & 2 \\ \mathbf{4} & -1 \\ \mathbf{1} & -3 \end{bmatrix} = \begin{bmatrix} (2)(1)+(1)(4)+(3)(1) \end{bmatrix}$

Continue to use the first row of A, this time going *further across* to align with the 2nd column of B. We similarly go *further across* to place our answer:

$\begin{bmatrix} \mathbf{2} & \mathbf{1} & \mathbf{3} \\ 0 & -1 & 2 \end{bmatrix} \begin{bmatrix} 1 & \mathbf{2} \\ 4 & \mathbf{-1} \\ 1 & \mathbf{-3} \end{bmatrix} = \begin{bmatrix} 9 & (2)(2)+(1)(-1)+(3)(-3) \end{bmatrix}$

Having 'exhausted' the 1st row of A we now move *down* to use the 2nd row and similarly move *down* to place our answer:

$\begin{bmatrix} 2 & 1 & 3 \\ \mathbf{0} & \mathbf{-1} & \mathbf{2} \end{bmatrix} \begin{bmatrix} \mathbf{1} & 2 \\ \mathbf{4} & -1 \\ \mathbf{1} & -3 \end{bmatrix} = \begin{bmatrix} 9 & -6 \\ (0)(1)+(-1)(4)+(2)(1) & \end{bmatrix}$

Continuing the process:

$\begin{bmatrix} 2 & 1 & 3 \\ \mathbf{0} & \mathbf{-1} & \mathbf{2} \end{bmatrix} \begin{bmatrix} 1 & \mathbf{2} \\ 4 & \mathbf{-1} \\ 1 & \mathbf{-3} \end{bmatrix} = \begin{bmatrix} 9 & -6 \\ -2 & (0)(2)+(-1)(-1)+(2)(-3) \end{bmatrix}$

Thus $\begin{bmatrix} 2 & 1 & 3 \\ 0 & -1 & 2 \end{bmatrix} \begin{bmatrix} 1 & 2 \\ 4 & -1 \\ 1 & -3 \end{bmatrix} = \begin{bmatrix} 9 & -6 \\ -2 & -5 \end{bmatrix}$

Confirm this result using your calculator.

Using your calculator to determine the product of matrices can be useful but if the matrices are not too big you should be able to determine the answers mentally. You would not need to show each step of the process and, with practice, you should be able to write the answer directly, as shown at the top of the next page.

$\begin{bmatrix} 2 & 1 & 3 \\ 0 & -1 & 2 \end{bmatrix} \times \begin{bmatrix} 1 & 2 \\ 4 & -1 \\ 1 & -3 \end{bmatrix}$

$\begin{bmatrix} 9 & -6 \\ -2 & -5 \end{bmatrix}$

ISBN 9780170390194

$$\begin{bmatrix} 1 & 3 \end{bmatrix}\begin{bmatrix} 2 & 1 \\ -1 & 4 \end{bmatrix} = \begin{bmatrix} -1 & 13 \end{bmatrix} \qquad \begin{bmatrix} 3 & 2 \\ 1 & 5 \end{bmatrix}\begin{bmatrix} 1 \\ 3 \end{bmatrix} = \begin{bmatrix} 9 \\ 16 \end{bmatrix}$$

$$\begin{bmatrix} 2 & 3 \\ -1 & 2 \end{bmatrix}\begin{bmatrix} 1 & 4 \\ 1 & 3 \end{bmatrix} = \begin{bmatrix} 5 & 17 \\ 1 & 2 \end{bmatrix} \qquad \begin{bmatrix} 3 & 1 \end{bmatrix}\begin{bmatrix} 1 & 2 & 1 \\ -1 & 0 & 1 \end{bmatrix} = \begin{bmatrix} 2 & 6 & 4 \end{bmatrix}$$

As was mentioned earlier, this method of matrix multiplication means that:

> Two matrices can be multiplied together provided the number of columns in the first matrix equals the number of rows in the second matrix.

Suppose matrix A has dimensions m × n and matrix B has dimensions p × q.

*   The product $A_{mn}B_{pq}$ can only be formed if n = p.
    In this case AB will have dimensions m × q.

*   The product $B_{pq}A_{mn}$ can only be formed if q = m.
    In this case BA will have dimensions p × n.

Note: In the product AB we say that B is *premultiplied* by A or that A is *postmutiplied* by B.

EXAMPLE 2

If $A = \begin{bmatrix} 1 & 2 & 3 \\ 1 & 4 & 0 \end{bmatrix}$, $B = \begin{bmatrix} 2 & 1 \\ 3 & -1 \end{bmatrix}$ and $C = \begin{bmatrix} 1 \\ 4 \\ 1 \end{bmatrix}$, determine each of the following.

If any cannot be determined state this clearly and explain why.

a   AB   b   BA   c   AC   d   CA   e B^2

Solution

a $AB = \begin{bmatrix} 1 & 2 & 3 \\ 1 & 4 & 0 \end{bmatrix}\begin{bmatrix} 2 & 1 \\ 3 & -1 \end{bmatrix}$ which cannot be determined because the number of columns in

A (**2 × 3**) ≠ the number of rows in B (**2 × 2**).

b $BA = \begin{bmatrix} 2 & 1 \\ 3 & -1 \end{bmatrix}\begin{bmatrix} 1 & 2 & 3 \\ 1 & 4 & 0 \end{bmatrix} = \begin{bmatrix} 3 & 8 & 6 \\ 2 & 2 & 9 \end{bmatrix}$ c $AC = \begin{bmatrix} 1 & 2 & 3 \\ 1 & 4 & 0 \end{bmatrix}\begin{bmatrix} 1 \\ 4 \\ 1 \end{bmatrix} = \begin{bmatrix} 12 \\ 17 \end{bmatrix}$

d $CA = \begin{bmatrix} 1 \\ 4 \\ 1 \end{bmatrix}\begin{bmatrix} 1 & 2 & 3 \\ 1 & 4 & 0 \end{bmatrix}$ which cannot be determined because the number of columns in

C (**3 × 1**) ≠ the number of rows in A (**2 × 3**).

e $B^2 = \begin{bmatrix} 2 & 1 \\ 3 & -1 \end{bmatrix}\begin{bmatrix} 2 & 1 \\ 3 & -1 \end{bmatrix} = \begin{bmatrix} 7 & 1 \\ 3 & 4 \end{bmatrix}$

Confirm the above answers using your calculator.

EXAMPLE 3

A manufacturer makes three products A, B and C, each requiring a certain number of units of commodities P, Q, R, S and T. Matrix X shows the number of units of each commodity required to make one of each product.

$$X = \begin{array}{c} \\ \text{Product A} \\ \text{Product B} \\ \text{Product C} \end{array} \begin{array}{ccccc} P & Q & R & S & T \\ \left[\begin{array}{ccccc} 1 & 1 & 0 & 2 & 3 \\ 1 & 1 & 2 & 1 & 2 \\ 0 & 1 & 3 & 0 & 3 \end{array}\right] \end{array}$$

a Each unit of P, Q, R, S and T costs the manufacturer $200, $100, $50, $400 and $300 respectively. Write this information as matrix Y which should be either a column matrix or a row matrix, whichever can form a product with X.

b Form the product referred to in **a** and explain what information it displays.

Solution

a As a column matrix, Y would have dimensions 5×1.
As a row matrix, Y would have dimensions 1×5.
Matrix $X_{3 \times 5}$ can form a product with $Y_{5 \times 1}$: $X_{3 \times 5}Y_{5 \times 1} = Z_{3 \times 1}$

$$\text{Thus } Y = \begin{bmatrix} \$200 \\ \$100 \\ \$50 \\ \$400 \\ \$300 \end{bmatrix} \begin{array}{l} \leftarrow \text{Cost of 1 unit of P} \\ \leftarrow \text{Cost of 1 unit of Q} \\ \leftarrow \text{Cost of 1 unit of R} \\ \leftarrow \text{Cost of 1 unit of S} \\ \leftarrow \text{Cost of 1 unit of T} \end{array}$$

(The order P, Q, R, S, T being consistent with the order in X.)

b
$$XY = \begin{bmatrix} 1 & 1 & 0 & 2 & 3 \\ 1 & 1 & 2 & 1 & 2 \\ 0 & 1 & 3 & 0 & 3 \end{bmatrix} \begin{bmatrix} 200 \\ 100 \\ 50 \\ 400 \\ 300 \end{bmatrix}$$

$$= \begin{bmatrix} 2000 \\ 1400 \\ 1150 \end{bmatrix} \begin{array}{l} \leftarrow \text{total commodity cost (\$) for producing 1 unit of product A} \\ \leftarrow \text{total commodity cost (\$) for producing 1 unit of product B} \\ \leftarrow \text{total commodity cost (\$) for producing 1 unit of product C} \end{array}$$

Note: While this chapter has considered

- adding and subtracting matrices,
- multiplying a matrix by a scalar

and • multiplying matrices

the concept of dividing one matrix by another is undefined for matrices.

Exercise 6B

Determine each of the following products. If any are not possible state this clearly and explain why.

1 $\begin{bmatrix} 1 & 2 \end{bmatrix} \begin{bmatrix} 2 & 3 \\ 1 & 3 \end{bmatrix}$

2 $\begin{bmatrix} 2 & 3 \\ 1 & 3 \end{bmatrix} \begin{bmatrix} 1 & 2 \end{bmatrix}$

3 $\begin{bmatrix} 2 & -1 \\ 1 & 0 \end{bmatrix} \begin{bmatrix} 1 & 4 \\ 0 & -2 \end{bmatrix}$

4 $\begin{bmatrix} 3 & 1 \end{bmatrix} \begin{bmatrix} 1 \\ 4 \end{bmatrix}$

5 $\begin{bmatrix} 1 \\ 4 \end{bmatrix} \begin{bmatrix} 3 & 1 \end{bmatrix}$

6 $\begin{bmatrix} 2 & -3 \\ -1 & 4 \end{bmatrix} \begin{bmatrix} 2 & 1 \\ -3 & 2 \end{bmatrix}$

7 $\begin{bmatrix} 1 & 0 \\ 0 & 1 \end{bmatrix} \begin{bmatrix} 2 & 3 \\ 1 & -1 \end{bmatrix}$

8 $\begin{bmatrix} 1 & 4 \\ -1 & 3 \end{bmatrix} \begin{bmatrix} 1 & 0 \\ 0 & 1 \end{bmatrix}$

9 $\begin{bmatrix} 0 & 0 \\ 0 & 0 \end{bmatrix} \begin{bmatrix} 2 & 1 \\ 4 & 5 \end{bmatrix}$

10 $\begin{bmatrix} 3 & 1 \\ 5 & 2 \end{bmatrix} \begin{bmatrix} 2 & -1 \\ -5 & 3 \end{bmatrix}$

11 $\begin{bmatrix} 8 & -5 \\ -3 & 2 \end{bmatrix} \begin{bmatrix} 2 & 5 \\ 3 & 8 \end{bmatrix}$

12 $\begin{bmatrix} 3 & 1 \\ 1 & 1 \end{bmatrix} \begin{bmatrix} 0.5 & -0.5 \\ -0.5 & 1.5 \end{bmatrix}$

13 $\begin{bmatrix} 1 & 2 & 1 & 2 \end{bmatrix} \begin{bmatrix} 2 \\ 1 \\ 2 \\ 1 \end{bmatrix}$

14 $\begin{bmatrix} 1 & 0 & 1 & 0 \\ 0 & 1 & 0 & 1 \end{bmatrix} \begin{bmatrix} 1 & 0 & 1 \\ 3 & -1 & 0 \\ 2 & 2 & 2 \\ 1 & 4 & 1 \end{bmatrix}$

15 $\begin{bmatrix} 1 & 0 \\ 0 & 2 \\ 1 & 1 \end{bmatrix} \begin{bmatrix} 1 & 0 & 5 \\ 5 & 1 & -1 \end{bmatrix}$

16 $\begin{bmatrix} 1 & 3 & 1 \\ 3 & 0 & -2 \end{bmatrix} \begin{bmatrix} 1 & 2 \\ 4 & 1 \\ -3 & -2 \end{bmatrix}$

17 $\begin{bmatrix} 1 & 2 & 3 \\ 4 & 5 & 6 \end{bmatrix} \begin{bmatrix} 1 \\ 2 \\ 3 \end{bmatrix}$

18 $\begin{bmatrix} 2 & 1 & 0 \\ -1 & 3 & 2 \\ 0 & 2 & 4 \end{bmatrix} \begin{bmatrix} 1 & 1 & -1 \\ 0 & 2 & 3 \\ 3 & 1 & 4 \end{bmatrix}$

19 If $A = \begin{bmatrix} 1 & 0 & -1 \\ 2 & 0 & 1 \\ 0 & 1 & 1 \end{bmatrix}$ and $B = \begin{bmatrix} 0 & 1 & 2 \\ 2 & 1 & 0 \\ 0 & -1 & 1 \end{bmatrix}$, determine the following:

a AB   **b** BA   **c** A^2 **d** B^2

20 Multiplication of numbers is commutative, i.e. if x and y represent numbers then xy is always equal to yx. Is matrix multiplication commutative for all pairs of matrices for which the necessary products can be formed? Justify your answer.

21 Provided the necessary products can be formed, matrix multiplication is associative, i.e. $(AB)C = A(BC)$.

Verify this for   **a** $A = \begin{bmatrix} 1 & 2 \\ -1 & 0 \end{bmatrix}, B = \begin{bmatrix} 3 & 1 \\ 0 & -1 \end{bmatrix}, C = \begin{bmatrix} 1 & 2 \\ -1 & 1 \end{bmatrix}.$

                              b $A = \begin{bmatrix} 1 & 2 \end{bmatrix}, B = \begin{bmatrix} 1 & 0 & -1 \\ 2 & 1 & 1 \end{bmatrix}, C = \begin{bmatrix} 1 & 0 \\ -1 & 2 \\ 1 & 1 \end{bmatrix}.$

22 Provided the necessary sums and products can be formed, the distributive law:
$$A(B + C) = AB + AC$$
holds for matrices.

Verify this for   **a** $A = \begin{bmatrix} 2 & 1 \\ 4 & 0 \end{bmatrix}, B = \begin{bmatrix} -1 & 1 \\ 0 & 1 \end{bmatrix}, C = \begin{bmatrix} 2 & 1 \\ -1 & 3 \end{bmatrix}.$

                              b $A = \begin{bmatrix} 2 & 0 \\ -3 & 1 \end{bmatrix}, B = \begin{bmatrix} 3 \\ 2 \end{bmatrix}, C = \begin{bmatrix} -1 \\ 4 \end{bmatrix}.$

23 If $A = \begin{bmatrix} a & b \\ c & d \end{bmatrix}$ and $B = \begin{bmatrix} e & f \\ g & h \end{bmatrix}$ and k is a number, prove that

$$(kA)B = A(kB) = k(AB)$$

24 A is a 3×2 matrix, B is a 3×2 matrix, C is a 2×3 matrix and D is a 1×3 matrix. State the dimensions of each of the following products. For any that cannot be formed state this clearly.

  a   AB   **b**   BA   **c**   BC   **d**   CB

  e   AD   **f**   DA   **g**   BCA   **h**   DAC

25 With $A = \begin{bmatrix} 2 & 1 \end{bmatrix}$, $B = \begin{bmatrix} 1 \\ 3 \end{bmatrix}$, $C = \begin{bmatrix} 1 & 4 \\ 2 & -1 \end{bmatrix}$ and $D = \begin{bmatrix} 1 \\ 3 \\ 1 \end{bmatrix}$ state whether each of the following products can be formed or not.

  a   AB   **b**   BA   **c**   AC   **d**   CA

  e   BD   **f**   DB   **g**   AD   **h**   DA

26 If it is possible to form the matrix product AA (i.e. A^2), what can we say about A?

27 BC is just one product that can be formed using two matrices selected from the three below. List all the other products that could be formed in this way. (The selection of the two matrices can involve the same matrix being selected twice.)

$$A = \begin{bmatrix} 1 & -1 \\ 2 & 1 \end{bmatrix} \qquad B = \begin{bmatrix} 1 & 3 \end{bmatrix} \qquad C = \begin{bmatrix} 4 \\ 1 \end{bmatrix}$$

28 a Premultiply $\begin{bmatrix} 2 & 0 \\ 3 & 2 \end{bmatrix}$ by $\begin{bmatrix} 1 & -1 \\ 2 & 0 \end{bmatrix}$.

b Postmultiply $\begin{bmatrix} 2 & 0 \\ 3 & 2 \end{bmatrix}$ by $\begin{bmatrix} 1 & -1 \\ 2 & 0 \end{bmatrix}$.

29 The 5×3 matrix shown on the right appeared on an earlier page. It shows the number of first, second and third places gained by each of five schools taking part in an inter-school sports carnival involving seven sports. Determine the rank order for these schools using the points matrix

	1st Place	2nd Place	3rd Place
School A	1	1	1
School B	3	1	0
School C	0	3	3
School D	1	2	0
School E	2	0	3

a 1st $\begin{bmatrix} 5 \text{ points} \\ 3 \text{ points} \\ 1 \text{ point} \end{bmatrix}$ 2nd, 3rd

b 1st $\begin{bmatrix} 4 \text{ points} \\ 3 \text{ points} \\ 2 \text{ points} \end{bmatrix}$ 2nd, 3rd

30 A financial adviser sets up share portfolios for three clients. Each portfolio involves shares in 4 companies with the number of shares as shown below.

	Abel Co.	Big Co.	Con Co.	Down Co.
Client 1	1000	5000	400	270
Client 2	500	8000	500	250
Client 3	500	3000	500	500

Initially the value of each share is:

Abel Co.	$5
Big Co.	50 cents
Con Co.	$12
Down Co.	$10

Two years later the value of each share is:

Abel Co.	$4
Big Co.	60 cents
Con Co.	$20
Down Co.	$10

Use matrix multiplication to determine the value of each client's portfolio at each of these times.

ISBN 9780170390194

31 A fast food outlet offers, amongst other things, Snack Packs and Family Packs. The contents of each of these are as in the contents matrix shown.

$$\begin{array}{c} \\ \text{Each Snack Pack} \\ \text{Each Family Pack} \end{array} \begin{array}{cc} \text{Drink (mL)} & \text{Number of burgers} \\ \left[\begin{array}{cc} 375 & 1 \\ 1250 & 4 \end{array} \right] \end{array}$$

An order comes in for 15 Snack Packs and 10 Family Packs. Use matrix multiplication to determine a matrix that shows the total volume of drink and the total number of burgers this order requires.

32 Three hotels each have single rooms, double rooms and suites. The number of each of these in each hotel is as shown in matrix P.

$$\begin{array}{c} \\ \text{Single} \\ \text{Double} \\ \text{Suite} \end{array} \begin{array}{ccc} \text{Hotel A} & \text{Hotel B} & \text{Hotel C} \\ \left[\begin{array}{ccc} 15 & 5 & 5 \\ 25 & 25 & 14 \\ 2 & 1 & 3 \end{array} \right] \end{array} = \text{Matrix P}$$

The three hotels are all owned by the same company and all operate the same pricing structure as shown in the tariff matrix, Q.

$$\begin{array}{c} \\ \text{Cost per night} \end{array} \begin{array}{ccc} \text{Single} & \text{Double} & \text{Suite} \\ \left[\begin{array}{ccc} \$75 & \$125 & \$180 \end{array} \right] \end{array} = \text{Matrix Q}$$

a Only one of PQ and QP can be formed. Which one?

b Determine the matrix product from part **a** and explain what information it is that this matrix displays.

c Suppose instead that the tariff matrix were written as a column matrix, R. The matrix product PR could be formed but would it give any useful data? Explain your answer.

33 A carpenter runs a business making three different models of cubby house for children. Each cubby house is made using four different sizes of treated pine timber. The number of metres of each size of timber required for each cubby house is shown below.

$$\begin{array}{c} \\ \\ \text{Cubby A} \\ \text{Cubby B} \\ \text{Cubby C} \end{array} \begin{array}{cccc} \text{Poles} & \text{Decking} & \text{Framing} & \text{Sheeting} \\ \text{120 mm diameter} & \text{90 mm} \times \text{22 mm} & \text{70 mm} \times \text{35 mm} & \text{120 mm} \times \text{12 mm} \\ \left[\begin{array}{cccc} 3 & 30 & 20 & 40 \\ 4 & 35 & 25 & 60 \\ 6 & 40 & 30 & 70 \end{array} \right] \end{array}$$

We will call this matrix P.

a The carpenter receives an order for 3 type As, 1 type B and 2 type Cs. Write this information as matrix Q which should be either a row matrix or a column matrix, whichever can form a product with P.

b Determine the product referred to in part **a** and explain what this matrix represents.

c The poles cost $4 per metre, the decking $2 per metre, the framing $3 per metre and the sheeting $1.50 per metre. Write this information as matrix R which should be either a row matrix or a column matrix, dependent on which will form a product with P. What dimensions would this product matrix be and what information would it display?

ISBN 9780170390194

34 A manufacturer makes four different models of a particular product. Matrix D below gives the number of units of commodities A, B and C required to make one of each model type.

	Model I	Model II	Model III	Model IV	
Commodity A	2	3	1	2	
Commodity B	20	30	50	40	= D
Commodity C	2	1	3	2	

a Each unit of the commodities A, B and C costs the manufacturer $800, $50 and $1000 respectively. Write this information as matrix E, either a column matrix or a row matrix, whichever can form a product with D.

b Form the product referred to in **a** and explain the information it displays.

35 A manufacturer makes three different models of a particular item. Matrix P below gives the number of minutes in the cutting area, the assembling area and the packing area required to make each model.

	Cutting	Assembling	Packing	
Each model A	30	20	10	
Each model B	20	30	10	= P
Each model C	40	40	10	

The manufacturer receives orders for 50 As, 100 Bs and 80 Cs.

We could write this as a column matrix, Q: $\begin{bmatrix} 50 \\ 100 \\ 80 \end{bmatrix}$

or as a row matrix, R: $\begin{bmatrix} 50 & 100 & 80 \end{bmatrix}$.

Both PQ and RP could be formed but only one of these will contain information likely to be useful.

a Which product is this?

b Form the product.

c Explain the information it gives.

Zero matrices

Any matrix which has every one of its entries as zero is called a zero matrix.

Thus the 2×2 zero matrix is $\begin{bmatrix} 0 & 0 \\ 0 & 0 \end{bmatrix}$.

The letter O is used to indicate a zero matrix. If it is necessary to indicate that it is the zero matrix of a particular dimension, say 2×2 for example, then we write $O_{2 \times 2}$.

There are obvious parallels between zero in the number system and a zero matrix in matrices.

6. Matrices ●●●●●●○○○○

With x representing a number:

$$x + 0 = x \qquad\qquad 0 + x = x \qquad\qquad x \times 0 = 0 \qquad\qquad 0 \times x = 0.$$

With A representing a matrix, and providing the necessary sums and products can be formed:

$$A + O = A \qquad\qquad O + A = A \qquad\qquad AO = O \qquad\qquad OA = O.$$

In this text we will use the letter O rather than the number 0 (zero) for the zero matrix. The two symbols can easily be confused, especially when handwritten. However this should not cause a problem, and you don't need to take time making some distinction between the handwritten characters, because the context usually makes it obvious whether it is the number zero or a zero matrix that is being referred to.

Because a zero matrix leaves another matrix 'unchanged under addition', ie. $A + O = A$ and also $O + A = A$, a zero matrix is sometimes referred to as an *additive identity matrix*.

Note: Care needs to be taken when working with matrices. We must guard against using rules and procedures that apply to numbers but that do not necessarily apply to matrices. For example, we have already seen that under matrix multiplication the matrix product AB is usually not the same as BA. Two more points to watch for are given below.

- With x and y representing numbers, a frequently used result in mathematics is that if $xy = 0$ then either $x = 0$ and/or $y = 0$.

  However, for matrices, if $AB = O$ it is not necessarily the case that A and/or $B = O$.

  For example consider $\quad A = \begin{bmatrix} 2 & 2 \\ 1 & 1 \end{bmatrix}$ and $B = \begin{bmatrix} 1 & -2 \\ -1 & 2 \end{bmatrix}$.

  In this case $\quad AB = \begin{bmatrix} 2 & 2 \\ 1 & 1 \end{bmatrix}\begin{bmatrix} 1 & -2 \\ -1 & 2 \end{bmatrix}$

  $$= \begin{bmatrix} 0 & 0 \\ 0 & 0 \end{bmatrix}$$

  Thus $AB = O$ but neither A nor B equal O.

- With x and y representing numbers, if $xy = zy$, for $y \neq 0$, then $x = z$.

  However, for matrices, if $AB = CB$, $B \neq O$, matrix A is not necessarily equal to matrix C. i.e. We cannot simply cancel the Bs.

  For example consider $\quad A = \begin{bmatrix} 3 & -1 \\ 1 & 4 \end{bmatrix}$, $B = \begin{bmatrix} 1 \\ 2 \end{bmatrix}$ and $C = \begin{bmatrix} -3 & 2 \\ 5 & 2 \end{bmatrix}$.

  In this case $\quad AB = \begin{bmatrix} 3 & -1 \\ 1 & 4 \end{bmatrix}\begin{bmatrix} 1 \\ 2 \end{bmatrix} = \begin{bmatrix} 1 \\ 9 \end{bmatrix}$

  and $\quad CB = \begin{bmatrix} -3 & 2 \\ 5 & 2 \end{bmatrix}\begin{bmatrix} 1 \\ 2 \end{bmatrix} = \begin{bmatrix} 1 \\ 9 \end{bmatrix}$.

  Thus $AB = CB$, $B \neq O$, but $A \neq C$.

ISBN 9780170390194

Multiplicative identity matrices

A multiplicative identity matrix leaves all other matrices unchanged under multiplication (provided the multiplication can be performed).

Thus if $I_{m \times n}$ is a multiplicative identity matrix then,

$$I_{m \times n} A_{n \times p} = A_{n \times p} \text{ from which it follows that } m = n.$$

Also $\quad B_{q \times m} I_{m \times n} = B_{q \times m}$ from which it follows that $m = n$.

Thus multiplicative identity matrices are square matrices.

The letter I is used to indicate a multiplicative identity matrix. If clarification is needed as to the size of I we can write I_2 for the 2×2 multiplicative identity, I_3 for the 3×3 multiplicative identity etc.

If we simply refer to an identity matrix it should be assumed that it is a multiplicative identity matrix, I, that is being referred to.

A multiplicative identity matrix has every entry of its main or leading diagonal equal to one and every other entry equal to zero.

The 2×2 multiplicative identity matrix is $\begin{bmatrix} 1 & 0 \\ 0 & 1 \end{bmatrix}$

The 3×3 multiplicative identity matrix is $\begin{bmatrix} 1 & 0 & 0 \\ 0 & 1 & 0 \\ 0 & 0 & 1 \end{bmatrix}$ etc.

$\begin{bmatrix} & & \\ & & \\ & & \end{bmatrix}$ *Main or leading diagonal*

There are obvious parallels between 1 in the number system and a multiplicative identity matrix in matrices.

With x representing a number: $\qquad\qquad 1 \times x = x, \qquad x \times 1 = x.$

With A representing a matrix: $\qquad\qquad I \times A = A, \qquad A \times I = A.$

Again be careful not to use rules and procedures that apply to numbers but that do not necessarily apply to matrices.

In numbers, if $\qquad\qquad\qquad\qquad xy = x, \quad$ then for $x \neq 0$, y must equal 1.

However, for matrices, if $\qquad\qquad AB = B, \quad B \neq O$, A does not necessarily equal I.

For example $\qquad\qquad\qquad \begin{bmatrix} 4 & -1 \\ -3 & 2 \end{bmatrix} \begin{bmatrix} 2 & 1 \\ 6 & 3 \end{bmatrix} = \begin{bmatrix} 2 & 1 \\ 6 & 3 \end{bmatrix}.$

Premultiplication by $\begin{bmatrix} 4 & -1 \\ -3 & 2 \end{bmatrix}$ has left $\begin{bmatrix} 2 & 1 \\ 6 & 3 \end{bmatrix}$ unchanged but $\begin{bmatrix} 4 & -1 \\ -3 & 2 \end{bmatrix} \neq I.$

Thus: Multiplication by I leaves a matrix unchanged, but a matrix being left unchanged under multiplication does not necessarily mean that it must have been I that we multiplied by.

That is, for matrices A and B, even if we know that AB = B we cannot assume that A = I, the identity matrix.

Route matrices

The diagram below shows the roads that exist between towns A, B, C and D.

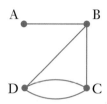

Notice how the following statements about the road network are written in the **route matrix** shown.

From A there is 1 direct road to B and none to C or D.
From B there is a direct road to A to C and to D.
From C there is no direct road to A, 1 to B and 2 to D.
From D there is no direct road to A, 1 to B and 2 to C.

$$
\text{From} \quad
\begin{array}{c}
 \\ A \\ B \\ C \\ D
\end{array}
\begin{array}{c}
 \\
\end{array}
\begin{array}{cccc}
A & B & C & D \\
\end{array}
\left[
\begin{array}{cccc}
0 & 1 & 0 & 0 \\
1 & 0 & 1 & 1 \\
0 & 1 & 0 & 2 \\
0 & 1 & 2 & 0
\end{array}
\right]
$$

To

Note: By *direct* road from A to B we mean a road from A to B that does not pass through any of the other towns in between.

The fact that all of the roads in the road system are 'two way' means that:

the number of roads from A to B is the same as from B to A,

the number from A to C is the same as the number from C to A,

the number from A to D is the same as the number from D to A, etc.

Can you see how this *symmetry* is also shown in the route matrix by the fact that
an entry in row m and column n
is the same as   the entry in row n and column m.

Note also that because there are no direct roads from
A back to A (without passing through any of B, C or D in between),
B back to B,
C back to C
or   D back to D
the leading diagonal of the route matrix contains only zeros.

The network below left has some one way roads and a road that goes directly from B back to B. The route matrix now does not have the same symmetry as the previous one and the leading diagonal is now not all zeros.

Note: The matrix shows two direct routes from B to B because the road loop can be travelled clockwise or anticlockwise.

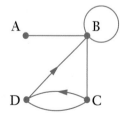

$$
\text{From} \quad
\begin{array}{c}
A \\ B \\ C \\ D
\end{array}
\begin{array}{cccc}
A & B & C & D \\
\end{array}
\left[
\begin{array}{cccc}
0 & 1 & 0 & 0 \\
1 & 2 & 1 & 0 \\
0 & 1 & 0 & 2 \\
0 & 1 & 1 & 0
\end{array}
\right]
$$

To

ISBN 9780170390194

Consider again the road system shown below.

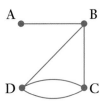

Notice that there is only one way of travelling from A to A *in two steps*: $\quad A \rightarrow B \rightarrow A$

and only one 'two stage' route from A to C: $\qquad\qquad\qquad\qquad A \rightarrow B \rightarrow C$

but there are two 'two stage' routes from B to C, as shown below:

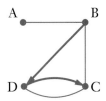 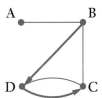

Confirm that continuing this thinking leads to the **two stage route matrix** shown below.

$$
\begin{array}{c}
 & & \text{To} \\
 & & \begin{array}{cccc} A & B & C & D \end{array} \\
\text{From} & \begin{array}{c} A \\ B \\ C \\ D \end{array} & \left[\begin{array}{cccc}
1 & 0 & 1 & 1 \\
0 & 3 & 2 & 2 \\
1 & 2 & 5 & 1 \\
1 & 2 & 1 & 5
\end{array}\right]
\end{array}
$$

Either mentally or with the aid of your calculator calculate the following matrix product,

$$
\begin{bmatrix}
0 & 1 & 0 & 0 \\
1 & 0 & 1 & 1 \\
0 & 1 & 0 & 2 \\
0 & 1 & 2 & 0
\end{bmatrix}
\begin{bmatrix}
0 & 1 & 0 & 0 \\
1 & 0 & 1 & 1 \\
0 & 1 & 0 & 2 \\
0 & 1 & 2 & 0
\end{bmatrix}
$$

i.e. (the one stage route matrix for this road system)^2, and compare your answer with the above two stage matrix.

Try to work out the two stage route matrix for the road network shown and then check your answer by determining, and then squaring, the one stage route matrix.

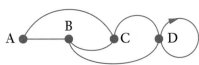

Note: We normally refer to a *one stage route matrix*, or *direct route matrix*, simply as a *route matrix*. If we mean the matrix showing the number of two stage routes or three stage routes we should call them *two stage route matrices* or *three stage route matrices*.

Social interaction as a matrix

We can use a matrix to indicate the existence, or otherwise, of some form of social interaction between members of a group.

Suppose the diagram on the right indicates who, in a group of five people, has visited the house of somebody else in that group. The diagram shows, for example,

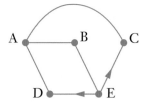

that   A has visited the house of B and of C and of D but has not visited the house of E,

         E has visited the house of D (and of B and C) but D has not visited the house of B, C or E,

         etc.

Using a '1' to indicate 'has visited' and a '0' to indicate 'has not visited' we can produce the matrix:

$$
\begin{array}{c}
\\ A \\ B \\ C \\ D \\ E
\end{array}
\begin{array}{ccccc}
A & B & C & D & E \\
\left[\begin{array}{ccccc}
0 & 1 & 1 & 1 & 0 \\
1 & 0 & 0 & 0 & 1 \\
1 & 0 & 0 & 0 & 0 \\
1 & 0 & 0 & 0 & 0 \\
0 & 1 & 1 & 1 & 0
\end{array}\right]
\end{array}
$$

Note: • In this situation we can reasonably assume that everyone has visited their own house, although these 'loops' are not included on the diagram. As this 'has visited your own house' is really rather pointless information we simply place zeros on the leading diagonal. On some occasions like this we might simply decide to put dashes on the leading diagonal. (However if we later wanted to multiply the matrix by a scalar or another matrix 'dashes' could cause a problem.)

       • In the above matrix we have used the convention that the matrix shows 'what the row person does to the column person'. For example the entry in row 5 (person E) and column 4 (person D) shows that person E has been to the house of person D.

Exercise 6C

1 If $A = \begin{bmatrix} 2 & 0 \\ -4 & -3 \end{bmatrix}$ find   **a** matrix B given that A + B = O, the 2 × 2 zero matrix.

                        b matrix C given that A + C = I, the 2 × 2 identity matrix.

2 If $D = \begin{bmatrix} 5 & -1 \\ 2 & 0 \end{bmatrix}$, O is the 2 × 2 zero matrix and I is the 2 × 2 identity matrix find

    a matrix E given that DO = E   **b** matrix F given that D + O = F

    c matrix G given that D + I = G   **d** matrix H given that DI = H

    e matrix J given that ID = J

3 For each of the following, state whether the given statement is necessarily true for all matrices A, B and C for which the given operations can be determined.

a AI = A

b IA = A

c AB = BA

d OA = O

e A + B = B + A

f A + A = 2A

g A(B + C) = AB + AC

h (AB)C = A(BC)

i If AB = O then A = O and/or B = O.

j If AB = AC and A ≠ O then B = C.

4 Determine the route matrix for each of the following road systems giving your answers in the form shown on the right.

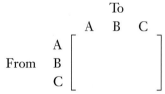

a 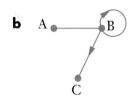   **b**   **c**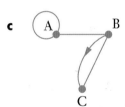

5 Determine the route matrix for each of the following road systems giving your answers in the form shown on the right.

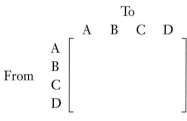

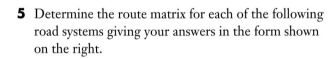

a 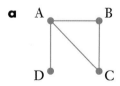   **b** 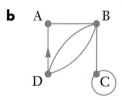   **c**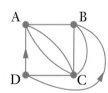

6 Determine the route matrix for each of the following road systems, giving your answers in the form shown.

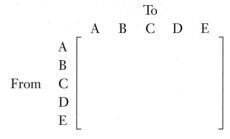

a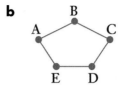   **b**   **c**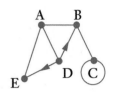

7 For each of the following, write down the two stage route matrix and then check your answer by calculating, and then squaring, the direct route matrix.

a   **b**

c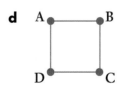   **d**

e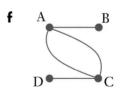   **f**

8 Investigate three stage route matrices.

9 A group of six people took part in a survey in which they were asked, in confidence, to list who of the other five people in the group they considered 'close friends'.

An analysis of the responses gave the diagram on the right.

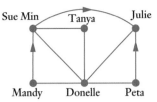

This diagram shows, for example that Donelle considers Julie a close friend and this feeling is reciprocated with Julie considering Donelle a close friend.

However, whilst Mandy considers Su Min a close friend this feeling is not reciprocated by Su Min, who does not consider Mandy a close friend.

These examples are also shown in the matrix below.

Placing zeros on the leading diagonal to make the 'being a close friend of oneself' count zero, copy and complete the matrix.

	Sue Min	Tanya	Julie	Peta	Donelle	Mandy
Sue Min						0
Tanya						
Julie					1	
Peta						
Donelle			1			
Mandy	1					

10 The diagram on the right shows 'who has been to the movies with who in a particular month' amongst a group of five friends, Ann, Bill, Chris, Dave and Enya.

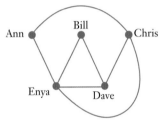

a Why are there no 'one way paths' in the diagram?

b What effect will this 'no one way paths' situation have on the matrix showing this 'been to the movies with each other' relationship?

c Construct the matrix of this situation using

- zeros to show a pairing has not been to the movies together in the month,
- ones to show that a pairing has been to the movies together in the month,
- zeros on the leading diagonal so that the situation of accompanying yourself to the movies is not counted.

d During a second month this group of five people were involved in three movie trips.

On one of these,   Ann, Bill and Enya went together,
on another,   Bill and Chris went together,
and   on another,   Dave went with Enya.

Construct a matrix like the one you did for part **c** for this second month.

11 Explain why it could be useful to be able to display social interactions using numbers in a matrix.

12 The diagram on the right shows 'who has whose phone number' in a group of seven school students.

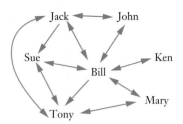

The arrows indicating, for example, that Jack has John's and John has Jack's but whilst Jack has Sue's, Sue does not have Jack's.

The matrix for this social interaction is started below with

- the usual convention observed in that the row person 'does the activity' to the column person. i.e. the cell a_{mn} shows whether or not m has the phone number of n,

- zeros used to indicate not having the number and ones used to indicate having the number,

- zeros placed in the leading diagonal so that the possession of one's own phone number will not count.

	Jack	John	Sue	Bill	Ken	Tony	Mary
Jack	0	1	1	1	0	1	0
John		0					
Sue			0				
Bill				0			
Ken					0		
Tony						0	
Mary							0

a Copy and complete the above matrix.

b Although Jack does not have Mary's number he could try to get it using one of the two-stage links   Jack → Bill → Mary
              Jack → Tony → Mary.

Copy and complete the following two stage matrix and then see how your answer compares with the matrix obtained by squaring your part **a** answer.

	Jack	John	Sue	Bill	Ken	Tony	Mary
Jack	0						2
John		0					
Sue			0				
Bill				0			
Ken					0		
Tony						0	
Mary							0

c Are there any of the seven who do not have the number of one of the others AND cannot obtain that number by a two stage process?

What feature of your two matrices shows this?

13 As part of the security system the top five agents in an undercover police operation can only initiate direct contact with some of the other four, but not all.

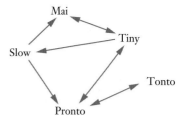

For example, as the diagram on the right shows, 'Slow' can initiate direct contact with 'Mai' and 'Pronto' but can only have direct contact with her initiated by 'Tiny'.

a Using the order as in the matrix below determine the one stage and two stage matrices for this situation with:

- the usual convention observed in that the row person 'does the activity' to the column person,

- placing one of just two possible entries in each space, either a zero (0) to indicate no contact or a one (1) to indicate contact,

- zeros placed in the leading diagonal on both the one stage and two stage matrices so that being able to contact oneself, either as a one step or two step process is not counted.

$$\begin{array}{c} \\ \text{Mai} \\ \text{Tiny} \\ \text{Tonto} \\ \text{Pronto} \\ \text{Slow} \end{array} \begin{array}{ccccc} \text{Mai} & \text{Tiny} & \text{Tonto} & \text{Pronto} & \text{Slow} \\ \left[\begin{array}{ccccc} & & & & \\ & & & & \\ & & & & \\ & & & & \\ & & & & \end{array} \right] \end{array}$$

b Is it the case that your two stage matrix is the square of your one stage matrix? If not, explain the differences.

Miscellaneous exercise six

This miscellaneous exercise may include questions involving the work of this chapter, the work of any previous chapters, and the ideas mentioned in the Preliminary work at the beginning of the book.

1 Matrix W_{mn} has m rows and n columns. For each of the following state whether the calculation can be performed. If it cannot be, then state this clearly, and if it can be, state the number of rows and columns the final matrix would have.

a $A_{23} + B_{32}$ **b** $2 \times A_{23}$ **c** $B_{32} - C_{25}$

d $B_{32} \times C_{25}$ **e** $D_{31} \times E_{31}$ **f** $E_{31} \times F_{14}$

g $(F_{14})^2$ **h** $(G_{33})^2$ **i** $H_{21} \times J_{21}$

j $J_{21} - K_{12}$ **k** $L_{12} \times M_{21}$ **l** $(N_{54} + P_{54}) \times R_{43}$

2 If $A = \begin{bmatrix} 5 & -2 \\ 3 & -1 \end{bmatrix}$ and $B = \begin{bmatrix} -1 & 2 \\ -3 & 5 \end{bmatrix}$ determine each of the following:

a $A + B$ **b** $2A$ **c** $A - B$

d AB **e** $(A - B)^2$ **f** $A^2 - B^2$

3 If a company does not have enough work to employ someone full time but would instead like to be able to call a worker in on the days they need them, the company may employ a worker as a *casual worker*. Casual work is usually temporary with no guarantee of it continuing long term and may involve irregular hours. A student working in a temporary position during college holidays would be a common example of casual work.

A company pays their permanent factory labourers $26.40 per hour. At certain times they need extra labourers and employ casual workers paying them 110% of the permanent rate, the higher rate reflecting the fact that casual workers have no paid sick leave, no long service leave and accumulate no holiday pay.

Calculate the weekly pay for each of the following people if the hours shown are for one week, permanent work force are paid time and a half for hours worked over 35 hours per week and casual workers receive the basic casual rate for all hours worked.

a Jaun, permanent, 37 hours   **b** Jackie, casual, 38 hours

c Su-Lin, permanent, 45 hours   **d** Ravinder, casual, 42 hours.

4 If an employee works *shiftwork* they can be *rostered* on to work their normal hours, e.g. 8 hours a day for 5 days a week, on any days from Monday to Sunday. This rostering could include some nights. Nursing, for example, would be one area where shiftwork would be commonly used.

Suppose a shift worker is paid $32.80 per hour day rate (DR). Night and weekend hourly rate (NWR) is 130% of the day rate and public holiday loading is 150% of either the DR or NWR as appropriate.

Calculate the weekly pay for each of the following:

a Sheila, 19 hours day duty + 8 hours night duty + 8 hours public holiday night duty.

b Tobias, 16 hours day duty + 16 hours night duty + 8 hours public holiday day duty.

5 Copy and complete the following table to compare various forms of interest for a loan of $25 000 at 9% per annum, using technology as you deem appropriate.

	\$25 000 borrowed at 9% per annum			
	Simple Interest	Compounded annually	Compounded every 6 months	Compounded quarterly
Initial amount borrowed				
Amount owed after 1 year				
Amount owed after 2 years				
Amount owed after 3 years				
Amount owed after 4 years				
Amount owed after 10 years				
Amount owed after 20 years				

7.

The theorem of Pythagoras

- Pythagoras' theorem
- Using the Pythagorean theorem to calculate the sides of right triangles
- Applications
- Miscellaneous exercise seven

Accurately draw a right triangle on a piece of A4 or A3 paper making the right triangle as large as you can whilst still leaving enough room to accurately draw a square on each side of the triangle, as shown below for triangle ABC.

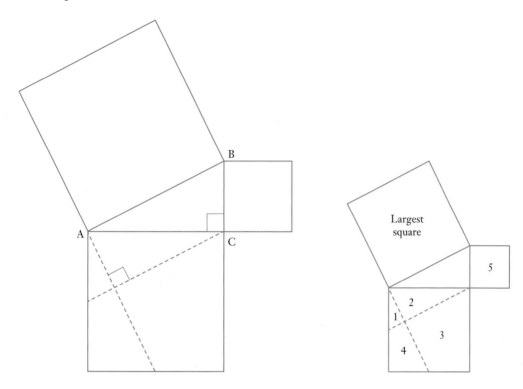

Now draw lines across the middle sized square, as shown by the broken lines in the diagram above. Cut out this square and accurately cut it up into the four pieces, labelled 1, 2, 3 and 4 in the smaller diagram. Also cut out the small square, labelled shape 5 in the smaller diagram. Try to fit shapes 1 to 5 together on the largest square to exactly fill it.

Pythagoras' theorem

Pythagoras was a Greek philosopher who lived approximately two and a half thousand years ago. One of the things he is most famous for is the result simply known as the theorem of Pythagoras (or Pythagoras' theorem or the Pythagorean theorem).

Note

- A *philosopher* is a person who seeks to understand, and gain knowledge about, the causes and underlying principles of things.
- A *theorem* is a statement, the truth of which can be established using existing truths.

If you successfully managed the activity on the previous page you demonstrated the truth of Pythagoras' theorem:

> The square of the length of the longest side of a right angled triangle is equal to the sum of the squares of the lengths of the other two sides.

Note: The longest side of a right triangle is the one opposite the right angle and is called the **hypotenuse**.

Thus, for the triangle shown on the right, $c^2 = a^2 + b^2$

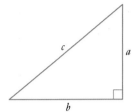

Using the Pythagorean theorem to calculate the sides of right triangles

The theorem of Pythagoras allows us to determine the length of one side of a right triangle, knowing the lengths of the other two sides, as the following examples demonstrate.

EXAMPLE 1

The right triangle sketched on the right has a hypotenuse of length x m. Find the value of x.

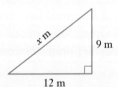

Solution

Using Pythagoras' theorem:
$$x^2 = 12^2 + 9^2$$
$$= 144 + 81$$
$$= 225$$

Solving gives $x = 15$ (Negative solution not applicable.)

Note:
- Asked to solve the equation $x^2 = 225$ there are *two* answers:
  $x = 15$ or -15 ($15^2 = 225$ and $(-15)^2 = 225$)
  However, in this context x m is the length of one side of a triangle so a negative value for x is not possible. Hence the statement in the above solution, *Negative solution not applicable.*

- Notice also that the question gives the length of the hypotenuse as x m. Hence x is a number, in this case a number of metres. In our answer we say $x = 15$. It would be incorrect to say $x = 15$ m as the length of the hypotenuse would then be 15 metre metres!

EXAMPLE 2

Find the length of side AC in the right triangle shown on the right, giving your answer in centimetres, correct to the nearest millimetre.

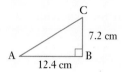

Solution

Using Pythagoras' theorem: $AC^2 = AB^2 + BC^2$

Thus if AC is of length x cm: $x^2 = 12.4^2 + 7.2^2$

$$= 205.6$$

Solving gives $x = 14.34$ correct to 2 decimal places

(Negative solution not applicable.)

Side AC is of length 14.3 cm, to the nearest millimetre.

EXAMPLE 3

Find the value of x in the diagram shown on the right, giving your answer correct to 1 decimal place.

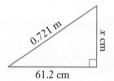

Solution

Using Pythagoras' theorem: $61.2^2 + x^2 = 72.1^2$

Thus $x^2 = 72.1^2 - 61.2^2$

$$= 1452.97$$

Solving gives $x \approx 38.1$ correct to 1 decimal place

(Negative solution not applicable.)

EXAMPLE 4

Find the value of x in the diagram shown on the right, giving your answer correct to one decimal place.

Solution

Let the height of the triangle be y cm (see diagram).

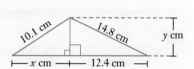

Applying Pythagoras to the further right of the two triangles:

$$14.8^2 = y^2 + 12.4^2$$

Hence $y^2 = 14.8^2 - 12.4^2$

$$= 65.28$$

i.e. $y = 8.1$ correct to 1 decimal place.

Now, from the triangle on the left of the two,

$$10.1^2 = y^2 + x^2$$

$$= 65.28 + x^2$$

Solving gives $x \approx 6.061$

$$= 6.1 \text{ correct to 1 decimal place.}$$

Note

Notice that the calculation for x used the accurate value for y, i.e. $\sqrt{65.28}$, rather than the 1 decimal place rounded value of 8.1 and thus avoided 'rounding errors'.

Exercise 7A

1 State the hypotenuse in each of the following triangles.

a

b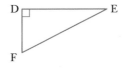

c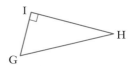

d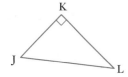

e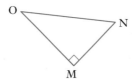

f

2 For the right triangle shown which of the following statements are true?
(There may be more than one correct statement.)

Statement I	Statement II	Statement III
$11^2 + x^2 = 5^2$	$x^2 + 11^2 = 5^2$	$11^2 + 5^2 = x^2$

Statement IV	Statement V	Statement VI
$5^2 + x^2 = 11^2$	$x^2 + 5^2 = 11^2$	$5^2 + 11^2 = x^2$

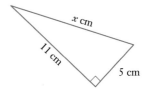

3 For the right triangle shown which of the following statements are true?
(There may be more than one correct statement.)

Statement I	Statement II	Statement III
$10^2 + 4^2 = x^2$	$10^2 + x^2 = 4^2$	$x^2 + 4^2 = 10^2$

Statement IV	Statement V	Statement VI
$x^2 = 4^2 - 10^2$	$x^2 = 10^2 - 4^2$	$4^2 = 10^2 - x^2$

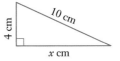

4 For the right triangle shown which of the following statements are true?
(There may be more than one correct statement.)

Statement I	Statement II	Statement III
$AC^2 = AB^2 + BC^2$	$AB^2 = AC^2 + CB^2$	$BC^2 = BA^2 + AC^2$

Statement IV	Statement V	Statement VI
$AC^2 = BC^2 - AB^2$	$BC^2 = AC^2 - AB^2$	$AB^2 = AC^2 - BC^2$

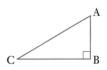

Find the value of x in each of the following, rounding your answers to 1 decimal place if rounding is necessary.

5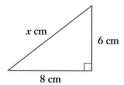

x cm, 6 cm, 8 cm

6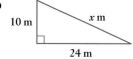

10 m, x m, 24 m

7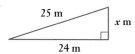

25 m, x m, 24 m

8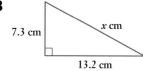

7.3 cm, x cm, 13.2 cm

9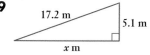

17.2 m, 5.1 m, x m

10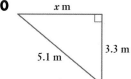

x m, 5.1 m, 3.3 m

11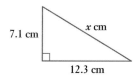

7.1 cm, x cm, 12.3 cm

12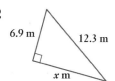

6.9 m, 12.3 m, x m

13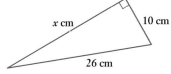

x cm, 10 cm, 26 cm

14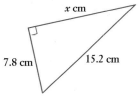

x cm, 7.8 cm, 15.2 cm

15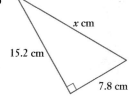

x cm, 15.2 cm, 7.8 cm

16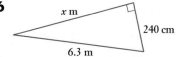

x m, 240 cm, 6.3 m

17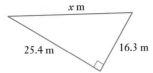

18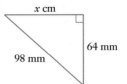

19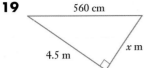

20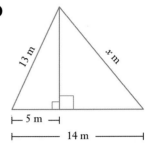

21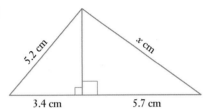

22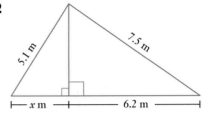

23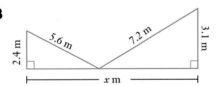

24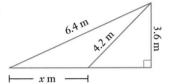

25 Triangle ABC is right angled at B and has AB = 26 mm and BC = 53 mm.
Find the length of AC, to the nearest millimetre.

26 Triangle PQR is right angled at R and has PQ = 17.3 cm and PR = 6.4 cm. Find the length of RQ,
to the nearest millimetre.

27 Triangle XYZ is a right triangle and has XY = 12.4 cm and YZ = 72 mm.
Find the possible lengths of XZ, to the nearest millimetre.

28 The diagram on the right shows the parallelogram
ABCD with all lengths as indicated.

By how much does the longer diagonal of the
parallelogram ABCD exceed the shorter diagonal?
(Answer to the nearest millimetre.)

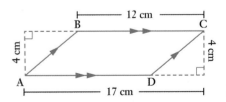

ISBN 9780170390194

Applications

The questions of the previous exercise all involved *abstract* right triangles - the triangles were not part of any practical situation. In the next examples, and the exercise that follows, the questions involve some real context. If such questions do not give a diagram of the situation you should draw one of your own to help you comprehend the question. Include given information on the diagram, introduce any letters you use and always make sure that your final statement does answer the question.

WS

Applications of
Pythagoras' theorem

EXAMPLE 5

A boat leaves a harbour, H, and travels 2.3 km due North and then 1.3 km due East. How far is the boat then from the harbour?

Solution

First draw a diagram and include on it any information that seems relevant.

Let the boat end up x km from the harbour.

By Pythagoras' theorem: $\quad x^2 = 2.3^2 + 1.3^2$

$\therefore \qquad\qquad\qquad\qquad x = \sqrt{2.3^2 + 1.3^2}$ (negative solution not applicable)

$\qquad\qquad\qquad\qquad\qquad \approx 2.64$

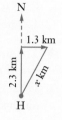

The boat is then 2.6 km from the harbour (correct to 1 decimal place).

Notice that the final answer was not simply $x = 2.64$. The letter x was not part of the initial problem, we introduced it to help obtain a solution. Interpreting the value of x back into the given situation, the final statement is given in terms of this situation and answers the question that was asked.

EXAMPLE 6

A ladder has its base on level ground and its top resting against a vertical wall. If the ladder is 5.1 metres in length and reaches 4.9 metres up the wall how far is the foot of the ladder from the base of the wall?

Solution

First draw a diagram and include on it any information that seems relevant.

Suppose that the foot of the ladder is x m from the base of the wall.

By Pythagoras' theorem: $\qquad\qquad\qquad\qquad x^2 + 4.9^2 = 5.1^2$

$\therefore \qquad\qquad\qquad\qquad\qquad\qquad x^2 = 5.1^2 - 4.9^2$

Solving and rejecting the negative solution gives $\qquad x \approx 1.41$

The foot of the ladder is approximately 1.4 metres from the base of the wall.

Exercise 7B

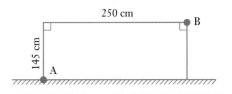

1 A show jumping fence is to be strengthened using a piece of timber from A to B, see diagram.

What length of timber is needed (to the nearest centimetre)?

2 A boat leaves a harbour and travels 5.63 km due West and then 1.32 km due North. How far is the boat then from the harbour? (Give your answer in km, rounded to one decimal place.)

3 A boat leaves a harbour and travels due East and then 2.12 km due South. The boat is then 3.54 km from the harbour. How far did the boat travel due East? (Give your answer in km, rounded to one decimal place.)

4 The size of a television screen is given as the size of the diagonal of the rectangular screen. What is the size of a rectangular screen with dimensions 56 cm by 42 cm?

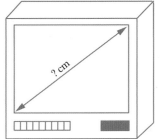

5 A ladder has its base on level ground and its top resting against a vertical wall. If the ladder is 8.5 metres in length and reaches 7.8 metres up the wall how far, to the nearest 10 cm, is the foot of the ladder from the base of the wall?

6 With its base on horizontal ground and 1.85 metres from a vertical wall a ladder just reaches the top of the wall. If the ladder is 5.25 metres in length find the height of the wall (to the nearest 10 centimetres).

7 A twenty metre mast is to be supported by four wires each with one end attached to the mast at a point two metres from the top and the other end attached to a point level with the base of the mast and five metres from it.

Rounding up to the next metre, what will be the length of each wire?

8 Ayetown lies 24 km due North of Ceetown.
Beetown lies 17 km due East of Ayetown.

A straight road links Ceetown to Ayetown and a straight road links Ayetown to Beetown.

A new straight road is planned to link Ceetown directly to Beetown.

Find how much this new road will reduce the journey from Ceetown to Beetown giving your answer in kilometres, correct to one decimal place.

ISBN 9780170390194

9 A rod of length 1 metre just fits inside a cylindrical container of base radius 25 cm.

Determine the height of the container, to the nearest cm.

10 Rather than use the footpath to travel around a rectangular park from point A to point B (see diagram) many people cut across the grass to travel in a straight line from A to B. How much shorter does this make the journey from A to B (to the nearest metre)?

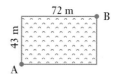

11 The diagram on the right shows a tent with all dimensions as indicated.

What is the area of each of the two rectangular sides of the tent?

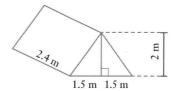

12 A company logo consists of a white square in a blue circle with each corner of the square touching the edge of the circle.

If the circle has a diameter of 12 cm what is the length of each side of the square, in millimetres, to the nearest millimetre?

13 A company is asked to quote a price for making 100 steel frames each in the shape of a right triangle with one side of length 1.2 metres and the hypotenuse of length 1.8 metres.

In order to make the quote the company needs to know, amongst other things, the length of steel required to make the 100 frames.

Calculate this total length rounding your answer up to the next whole metre.

14 Fencing is to be placed around a play area that is in the shape of the trapezium ABCD shown in the diagram on the right.

With lengths as shown in the diagram what is the perimeter of the trapezium? (Give your answer in metres correct to one decimal place.)

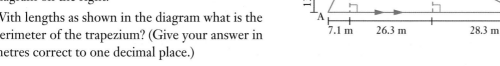

15 A crane has the basic structure shown on the right with all lengths as indicated.

Find the length of BC in metres, correct to one decimal place.

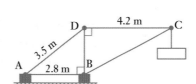

16 A watering system is to cover a square area of side length 8 metres with five sprinklers, one at each corner of the square and one at the centre. Which of the following systems of pipes uses the smaller total length of piping and how much smaller, to the nearest centimetre?

Layout A

Layout B

17 An area of land is in the shape of a right triangle with its hypotenuse of length 243.32 metres and one side of length 72.14 metres. Determine the area of this triangle, giving your answer to the nearest ten square metres.

Remember: the area of a triangle is $\dfrac{\text{base} \times \text{perpendicular height}}{2}$

18 A rectangular frame measuring 3 metres by 1.4 metres is to be made of steel rods. Which of the two designs shown below requires the greater length of steel and by how much (to the nearest centimetre)?

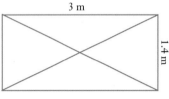

Design A

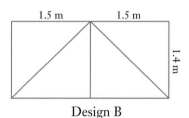

Design B

Suppose the frame was 1.8 metres by 1.4 metres instead.

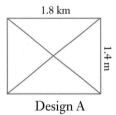

Design A

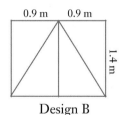

Design B

Now which design requires the greater length of steel and by how much? (Again give your answer to the nearest centimetre).

RESEARCH

Investigate Pythagorean triples

3 ✳ 4 ✳ 5   5 ✳ 12 ✳ 13

7 ✳ 24 ✳ 25   8 ✳ 15 ✳ 17

Miscellaneous exercise seven

This miscellaneous exercise may include questions involving the work of this chapter, the work of any previous chapters, and the ideas mentioned in the Preliminary work at the beginning of the book.

1 Increase $1000 by 15% and then decrease your answer by 15%.

2 a Premultiply $\begin{bmatrix} 1 & -2 \end{bmatrix}$ by $\begin{bmatrix} 2 \\ 3 \end{bmatrix}$.

  b Postmultiply $\begin{bmatrix} 1 & -2 \end{bmatrix}$ by $\begin{bmatrix} 2 \\ 3 \end{bmatrix}$.

3 a Increase $409.50 by 30%.
  b When an initial amount is increased by 30% it becomes $409.50. What was the initial amount?
  c Decrease $409.50 by 30%.
  d When an initial amount is decreased by 30% it becomes $409.50. What was the initial amount?
  e Find 30% of $409.50.

4 The matrices A, B and C shown below can be multiplied together to form a single matrix if A, B and C are placed in an appropriate order. What is the order and what is the single matrix this order produces?

$$A = \begin{bmatrix} 2 & 1 & 3 \\ 0 & -1 & 2 \end{bmatrix}, \quad B = \begin{bmatrix} 1 & -1 \end{bmatrix}, \quad C = \begin{bmatrix} 1 & 1 & 0 & -1 \\ 0 & 1 & -1 & 3 \\ 3 & 1 & 4 & 0 \end{bmatrix}.$$

5 What is the longest pole that could fit into the rectangular container shown on the right? (Give your answer to the nearest cm.)

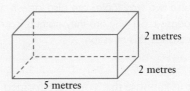

2 metres
2 metres
5 metres

6 The diagram below shows a vertical ladder leading to a water slide.

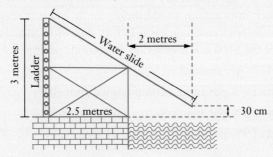

3 metres
Ladder
Water slide
2 metres
2.5 metres
30 cm

With all measurements as shown in the diagram determine the length of the water slide giving your answer to the nearest centimetre.

7 The route matrix on the right is for a road system between five points A, B, C, D and E. Explain how you can tell that the only 'one way roads' are from A to C and from E to D.

$$
\begin{array}{c} & & \text{To} \\ & & \begin{array}{ccccc} A & B & C & D & E \end{array} \\ \text{From} & \begin{array}{c} A \\ B \\ C \\ D \\ E \end{array} & \begin{bmatrix} 0 & 1 & 1 & 2 & 0 \\ 1 & 0 & 1 & 3 & 0 \\ 0 & 1 & 0 & 1 & 1 \\ 2 & 3 & 1 & 0 & 0 \\ 0 & 0 & 1 & 1 & 0 \end{bmatrix} \end{array}
$$

8 One common form of matrix display is the *two way classification table*, an example of which is shown below. This table classifies the 1237 members of a health club according to two categories, gender, i.e. male or female, and age, 'under 30' or '30 or over'.

	Under 30	30 or over	Totals
Male	329	276	605
Female	414	218	632
Totals	743	494	1237

To check that you understand the table, check that you agree with the following statements:

743 of the health club members were under 30.

There were 27 more female members than male members.

a How many male club members are aged 30 or over?

b To the nearest percent what percentage of the female members were under 30?

c To the nearest percent what percentage of the club members were males under 30?

> **Note**
>
> These two way tables are also known as Carroll diagrams, named after the mathematician Charles Dodgson. So why are they named Carroll diagrams and not Dodgson diagrams? Do some research to answer this question.

9 Let us suppose that you are informed by email that a distant relative of yours passed away many years ago and it has just been realised that during her life she invested $100 into an account and now, 150 years later, you are the beneficiary of the account plus interest accrued. Now whilst most people would assume such information would simply be 'an internet scam' let us suppose that in this case this most unlikely circumstance was in fact true. How much would this account plus interest now be worth if the $100 was invested in an account paying

a simple interest of 15% per annum?

b compound interest of 8% per annum compounded annually?

c compound interest of 8% per annum compounded quarterly?

d compound interest of 9% per annum compounded annually?

ISBN 9780170390194

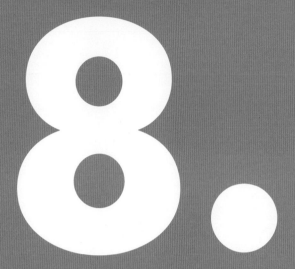

8.

Perimeter
and area

- Rectangles, parallelograms, triangles, trapeziums and circles
- Applications
- Inverse questions
- Miscellaneous exercise eight

Rectangles, parallelograms, triangles, trapeziums and circles

The *Preliminary work* section at the beginning of this book reminded you how to find the areas of rectangles, parallelograms, triangles, trapeziums and circles. With this knowledge you should also be able to determine the areas of shapes that are combinations of these shapes or that are parts of such shapes, as the following examples demonstrate.

Units of length and perimeter

Areas of composite shapes

A page of circular shapes

Area ID

EXAMPLE 1

Find the area of the shaded shape shown on the right.

Solution

Area of rectangle $= 16 \text{ cm} \times 6 \text{ cm}$

$= 96 \text{ cm}^2$

Area of triangle $= \dfrac{16 \text{ cm}}{2} \times 7 \text{ cm}$

$= 56 \text{ cm}^2$

Area of circle $= \pi \times (4 \text{ cm})^2$

$= 16\pi \text{ cm}^2$

Thus, shaded area $= 96 \text{ cm}^2 + 56 \text{ cm}^2 - 16\pi \text{ cm}^2$

$= 101.7 \text{ cm}^2$ (rounded to 1 decimal place)

EXAMPLE 2

Find the area and the perimeter of the shaded shape shown on the right.

Solution

Perimeter of shaded shape $= 4 \times 6 \text{ m} + \dfrac{2 \times \pi \times 6 \text{ m}}{360} \times 40$

$= 28.19 \text{ m}$ (to nearest cm).

Area of shaded shape $= 6 \text{ m} \times 6 \text{ m} + \dfrac{\pi \times (6 \text{ m})^2}{360} \times 40$

$= 48.57 \text{ m}^2$ (rounded to 2 decimal places)

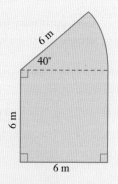

Exercise 8A

For questions 1 to 8 find **a** the perimeter and **b** the area of the shaded region.

1

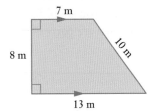

2

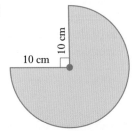

Give perimeter to the nearest mm and area in cm² to nearest cm².

3

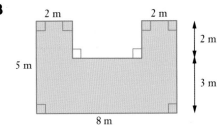

4

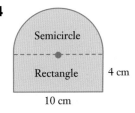

Give perimeter to the nearest mm and area in cm² to nearest cm².

5

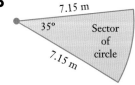

Give perimeter to the nearest cm and area in m² rounded to 2 decimal places.

6

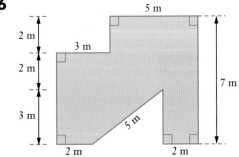

7

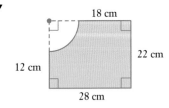

Give perimeter to the nearest mm and area in cm² to nearest cm².

8

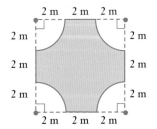

Give perimeter to the nearest cm and area in m² rounded to 2 decimal places.

ISBN 9780170390194

For questions 9 to 16 find the area of the shaded region.

9

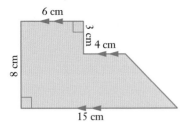

10

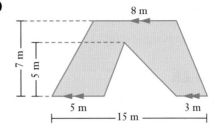

11

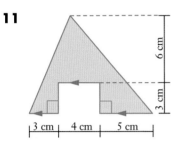

12

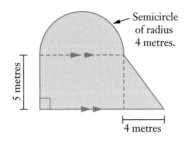

Semicircle of radius 4 metres.

Give area in m² rounded to 1 decimal place.

13

The circles each have a diameter of 1 metre.

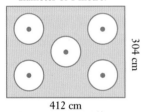

Give area to nearest cm².

14

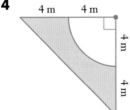

Give area in m² rounded to 2 decimal places.

15

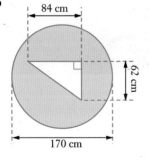

Give area to nearest cm².

16

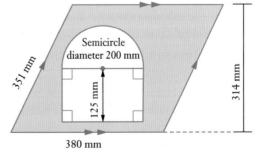

Semicircle diameter 200 mm

Give area to nearest 10 mm².

Applications

In the following exercise, you will again need to determine the perimeter and area of various shapes. However, the questions now involve some everyday context for which determining the perimeter or area of a shape is significant.

Exercise 8B

1 The diagram below shows Bernard's block of land and its various different areas.

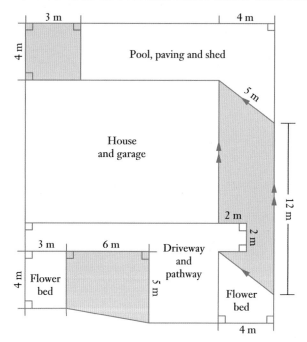

a For the three shaded areas Bernard plans to lay lawn using 'roll on real grass' which he can purchase for $11.20 per square metre.

To make sure he has enough lawn Bernard plans to calculate the total area and then round this total up to the next multiple of 5 square metres.

How much will the order for 'roll on real grass' cost Bernard?

b Alternatively, Bernard could have synthetic grass in these areas which he could buy for $24 per square metre and lay it himself, or for $37 per square metre he could buy it and have it laid professionally. Again rounding up to the next multiple of 5 m² how much would each of these options cost?

c Bernard is considering having concrete edging put along all of the lawn area edges that do not form edges with the house and garage area, driveways, pathways, or the property boundary.

Bernard can get this edging done for $27 per metre.

How much would this edging cost him?

2 'Skirting board' is the decorative strip of wood that runs around the base of the interior walls of a house to cover any gap and uneven edges where the wall meets the floor.

Ignoring any door spaces or other places where skirting board would not be placed, a room with the rectangular floor shape shown on the right, would require 15.8 metres of skirting board (= 4.4 m + 3.5 m + 4.4 m + 3.5 m).

Suppose that skirting board is sold in lengths of 2.4 metres and 3 metres, and that only complete lengths can be purchased.

For the room illustrated we could buy:

     7 lengths of 2.4 metres   and have   1 metre excess,
or   6 lengths of 3 metres   and have   2 metres excess,
or, for minimum excess:
3 lengths of 3 metres and 3 lengths of 2.4 metres and have just 0.4 metres excess.

For each of the floor shapes shown below, determine statements like those shown for the 4.4 m by 3.5 m floor plan above.

I.e. One statement giving the number of 2.4 metre lengths plus excess.
     One statement giving the number of 3 metre lengths plus excess.

And also

     One statement giving the combination of the two lengths that will give the minimum excess. (Challenging.)

* Ignore any door spaces etc where skirting board would not be placed.

* Assume that the skirting board can be purchased in just two standard lengths, 2.4 metres and 3 metres.

* Assume that only complete lengths can be purchased.

a

b

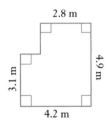

c

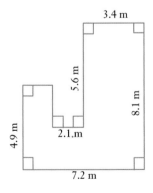

> ### Note
>
> In reality, some people would choose the lengths to minimise the number of joins, door spaces would not be ignored, some excess would be wanted to allow for joins etc. and some other lengths may well be available for purchase.

---

3 A newspaper offers advertising space in five standard sizes, A to E, as indicated below (not drawn to full size).

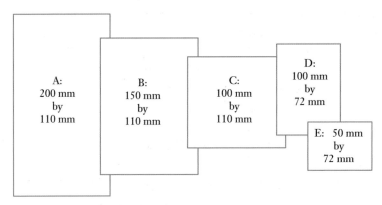

A:
200 mm
by
110 mm

B:
150 mm
by
110 mm

C:
100 mm
by
110 mm

D:
100 mm
by
72 mm

E: 50 mm
by
72 mm

For each size the cost for an advertisement has a basic price of $18.50 per square centimetre. The paper keeps some pages free of any advertisements and for other pages it has the following loadings added if particular pages are requested:

Page in paper	Percentage added to basic price
Page 3	60%
Page 5	45%
Page 7, 9 or 11	30%
Any other requested page	10%

If no particular page is requested the editors will place the advertisement on the page that best suits the newspaper lay out and there will be no percentage loading made to the basic price.

Find the cost of each of the following advertisements with this newspaper.

	Size	Page	Requested or Not requested
a	A	5	Requested
b	E	7	Requested
c	B	14	Not requested
d	D	3	Requested
e	D	7	Requested
f	C	13	Requested
g	A	7	Requested
h	E	5	Not requested

ISBN 9780170390194

4 A company specialising in applying a protective floor covering to factory floors, warehouse floors, garage floors etc has a special offer, as shown in the advertisement on the right.

The company calculates the area of the floor involved, charges $45/m², reduces this price by 12% during the special offer and then rounds down to whole numbers of dollars.

Under this special offer what would this company charge for applying the industrial strength floor covering to each of the floor areas shown shaded below?

Note: Angles that look right angled in the diagram should be assumed to be right angles.

a

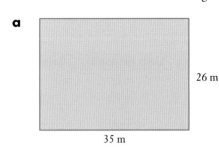

b

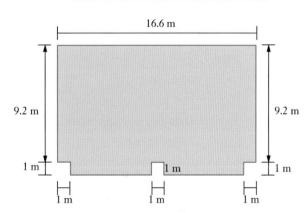

c

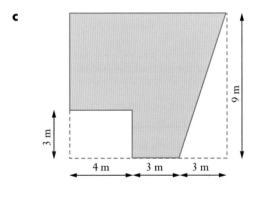

d

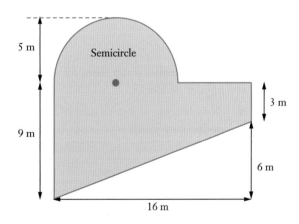

5 The diagram below shows a section of a particular type of farm fencing, and gate.

The fencing consists of horizontal wooden fence rails each of 3.6 metres in length, attached to wooden fence posts which are cemented into the ground at 1.8 metre intervals. The gate can be purchased ready made and, together with its posts, spans the gap between posts as shown below.

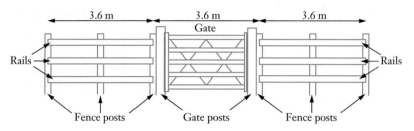

Lengths that are not exact multiples of 3.6 metres can be fenced by cutting rails to an appropriate length. The order form/invoice for the materials required for the section shown in the diagram above would be as shown below.

Qty	Item	Unit price ($)	Total ($)
6	Fence post	10.00	60.00
6	Post pack (Cement, brackets and nails for 1 post)	18.00	108.00
1	Gate pack (Gate, gate posts, latch, cement and all fastenings)	240.00	240.00
6	3.6 metre rail	8.00	48.00
		Sub total	$456.00
		GST (10%)	$45.60
		Grand total	$501.60

Complete a similar order form for using this style of fencing for each of the following rectangular paddocks.

a New fencing along AB and BC with a gate in AB.

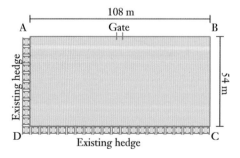

b Fence entire perimeter of rectangle EFGH with a gate in HG.

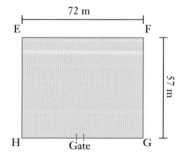

ISBN 9780170390194

6 For a '6-week picking time' the owner of a mushroom farm expects his farm to produce an average yield of 16 kg of mushrooms for each square metre under cultivation.

What quantity of mushrooms should the farmer expect in one such 6-week picking time from a system which involves a cultivated area consisting of 18 sets of the 5 tray rectangular stacking structure shown below (not drawn to scale)?

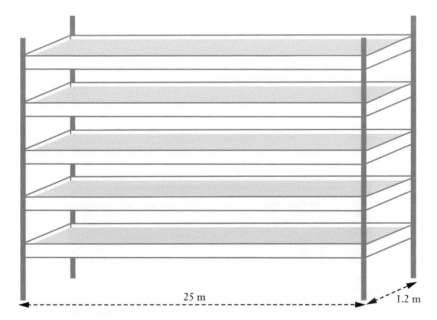

25 m

1.2 m

7 A city council is investigating the cost of constructing some multi-storey car parks reasonably close to a proposed sports and entertainment centre. Research suggests that 325 parking spaces can be situated on each hectare (100 m × 100 m) of available space. Thus a high rise parking facility with 5 floors and 0.4 hectares of available space on each floor could accommodate 650 parking spaces.

Research also suggests that the construction costs for the sort of structures the council is considering cost, on average, $22 000 per parking space, once the costs of access, lifts, ramps, lighting, stairs etc have been included.

The council is also advised that once a multi-story car park is built the ongoing maintenance costs to cover, lighting, repairs, security, fee collection etc average $450 per parking space per year. (While this should be more than covered by the parking fees council want to know what the cost is likely to be for planning purposes.)

Based on these figures, what would be the cost of constructing multi-storey car parks for which each floor has the available space shown and the number of floors (including the ground floor) is as indicated? Also determine the annual ongoing maintenance costs for each car park. (Angles that appear to be right angles on the diagrams should be assumed to be right angles.)

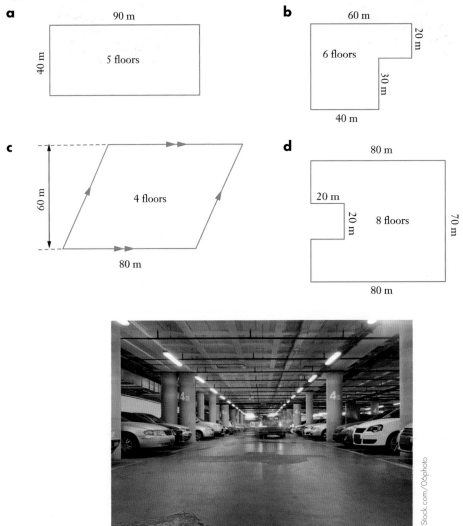

a 90 m, 40 m, 5 floors

b 60 m, 20 m, 30 m, 40 m, 6 floors

c 60 m, 80 m, 4 floors

d 80 m, 20 m, 20 m, 8 floors, 70 m, 80 m

8 In a particular wheat growing region farmers can expect that, with the usual weather pattern, if they sow 80 kg of seed per hectare (100 m × 100 m) they will achieve a wheat yield of approximately 180 grams per square metre provided they properly prepare and maintain the land.

For this growing region, and based on the figures above, what quantity of seed would each of the following shaded areas require, to the nearest 10 kg, and how much wheat should a farmer expect to harvest from each area (given usual weather patterns and properly prepared and maintained land)?

Angles that appear right angled in the diagram should be assumed to be right angles.

pixabay.com/Hans

a

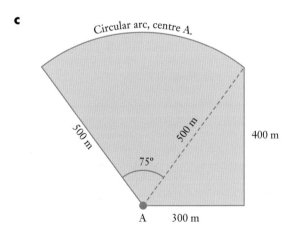

b

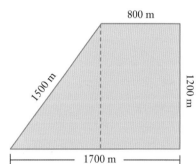

c

Circular arc, centre A.

d

Circular arc, centre B.

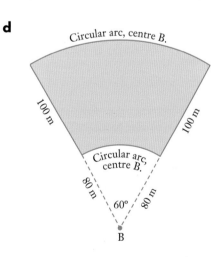

9 For the rate mentioned in the advertisement below find the price for having this company clean carpets that cover each of the shaded areas shown below. (Round each answer to the nearest whole dollar.)

Angles that look like right angles in the diagrams should be assumed to be right angles.

a

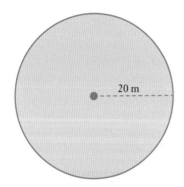

20 m

b

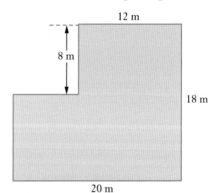

12 m

8 m

18 m

20 m

c

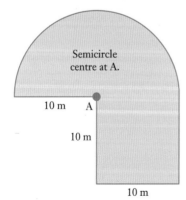

Semicircle centre at A.

10 m   A

10 m

10 m

d

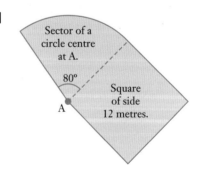

Sector of a circle centre at A.

80°

Square of side 12 metres.

A

10 Hair density of 80 follicular units per square centimetre is considered normal average hair density.

A hair transplant specialist tends to graft 50 follicular units per square centimetre onto bald areas because, although being less than the normal average hair density, the specialist considers the 50 units/cm^2 to be quite sufficient for the graft to appear to be normal density.

At this 50 follicular units/cm^2, how many follicular units will the specialist graft into each of the following bald areas. (Give answers rounded up to the next multiple of 10 follicular units.)

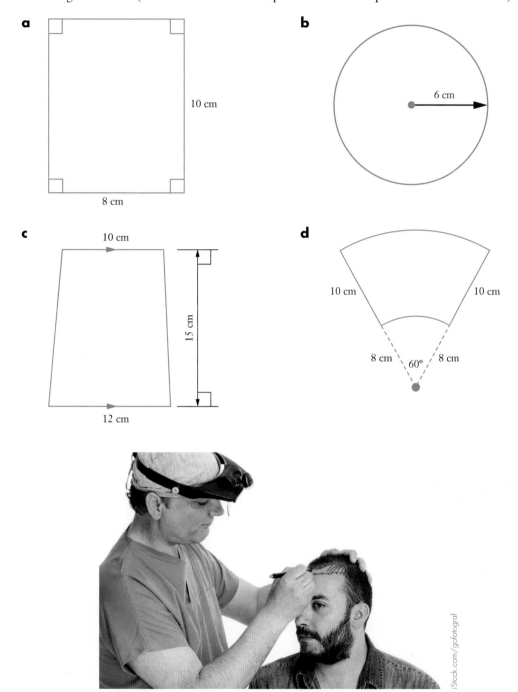

a

10 cm

8 cm

b

6 cm

c

10 cm

15 cm

12 cm

d

10 cm   10 cm

8 cm   60°   8 cm

iStock.com/gofotograf

11 A kitchen renovation company offers two types of stone for stone benchtops. The company calculates the price on a cost per square centimetre basis and then rounds to the nearest dollar to give the quote.

| Type: | Reconstituted | Cost: | $0.0384 per cm^2 |
| Type: | Natural | Cost: | $0.0435 per cm^2 |

• No charge is made for cutting out sink or stove spaces but the area is calculated as if these bits were not removed. (The cut out pieces have their edges polished and are given to the customers for use as chopping boards.)

• Any circular shapes or parts of circles will be charged on their area but then a 30% loading will be added to the cost of the circular area due to wastage.

• All exposed ends and edges are polished without any additional charge.

• Joins will be kept to a minimum but some will usually be necessary.

• 40 mm 'bull nose' front edges can be added for an additional cost of $1.00 per linear centimetre.

a Why does the cost of the bull nose edging refer to a cost per *linear* metre?

b What price will be quoted for each of the following stone bench tops?

Edges that are to have the 'bull nose' front edges added are shown below by the bolder lines. (Angles that look right angled in the diagram should be assumed to be right angled.)

i

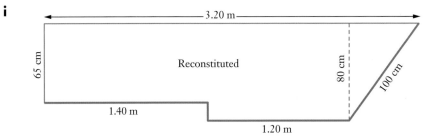

ii

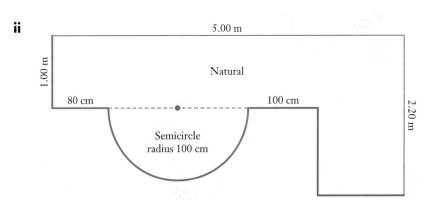

12 A hardware shop sells the *Brush It* brand of quality paint in various sizes of tin.

A tin containing	1 litre of the paint costs	$42.00
A tin containing	2 litres of the paint costs	$59.90
A tin containing	4 litres of the paint costs	$78.40
A tin containing	10 litres of the paint costs	$179.00

The manufacturer of the paint claims that one litre of paint will put one coat of paint on 16 m^2 of internal wall.

Assuming the manufacturer's claim is correct how many of each size tin of *Brush It* should be purchased to put **three** coats of paint on **all** of the following shaded wall areas and what will this paint cost?

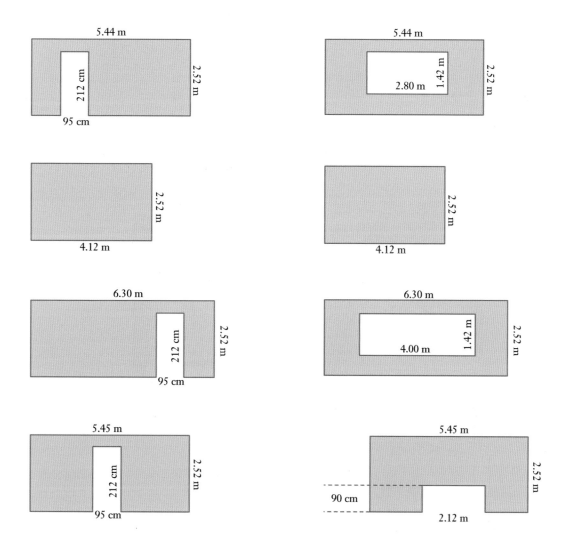

13 A company is developing a number of housing estates in various parts of Australia. Three projects, namely Maynard Waters, Plymptain by Sea and Woodstock Valley, have some identical blocks which, because of location, are priced differently. In these three estates the basic price of the land per square metre is as follows:

Maynard Waters	$485 per square metre
Plymptain by Sea	$520 per square metre
Woodstock Valley	$625 per square metre.

In each estate, each block is also graded as standard, premium or prime and these categories then have the above prices increased by 0%, 5% and 15% respectively.

Find the price of each of the following blocks in each of the three estates, giving your answers to the nearest $1000.

a

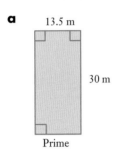

13.5 m
30 m
Prime

b

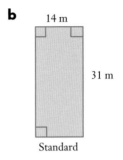

14 m
31 m
Standard

c

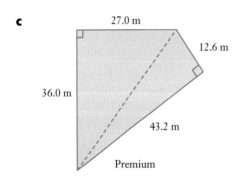

27.0 m
12.6 m
36.0 m
43.2 m
Premium

d

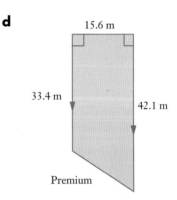

15.6 m
33.4 m
42.1 m
Premium

e

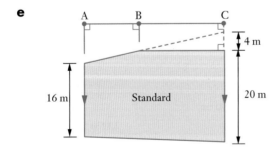

A   B   C
4 m
16 m   Standard   20 m

AC = 32 metres
AB : BC = 3 : 5

14 The cost of having a house built depends on the location, the amount of work required to prepare the site for building, the type of construction, the level of finish, the building company and the size of the house.

For a particular building company and region, and with the site ready for building, the following table gives an estimate of the construction cost per square metre of floor plan for a single level dwelling of various types and various finishes

Construction type	Finish		
	Basic	Basic Plus	Deluxe
3 bedroom brick veneer standard design	$1080	$1270	$1620
3 bedroom full brick unique design	$1420	$1670	$2130
4 bedroom brick veneer standard design	$1210	$1420	$1810
4 bedroom full brick unique design	$1590	$1870	$2380

Based on the figures in the table, create a similar table for a house with the floor plan as shown below (all measurements are in millimetres), but with your table showing the estimated price for building the house to each type and each finish. (Assume angles in the diagram that appear right angled are indeed right angled.)

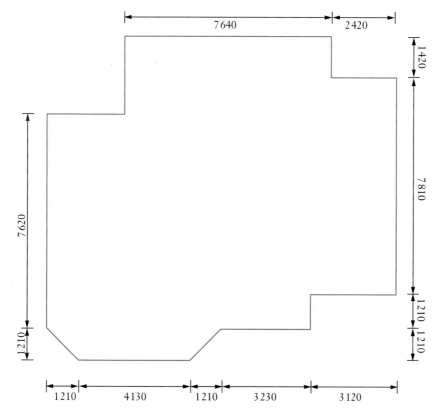

15 For general paving, rather than specialist or ornate work, a company charges $38.20/m^2$ for site preparation plus an additional $18.40/m^2$ if existing concrete work needs to be removed. On top of these site preparation costs the company charges $28.00/m^2$ for laying new pavers. The cost of the pavers is additional to this and the various types of pavers offered by the company are priced as follows:

Sandstone pavers	$28.95/m^2$	Limestone pavers	$37.95/m^2$
Bluestone pavers	$42.95/m^2$	Granite pavers	$49.95/m^2$

If the customer wishes, and once the pavers are laid, a sealant coating can be applied for a cost of $8.50/m^2$

Find the cost of paving each of the areas below using this company to do the paving, with the work and pavers required as indicated. (Angles that look like right angles in the diagram should be assumed to be right angles.)

a

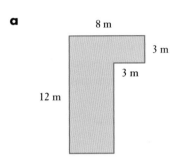

Site prep needed: Yes
Concrete removal needed: No
Sealant coating requested: No
Paver type: Sandstone

b

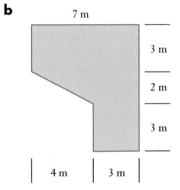

Site prep needed: Yes
Concrete removal needed: Yes
Sealant coating requested: Yes
Paver type: Bluestone

c

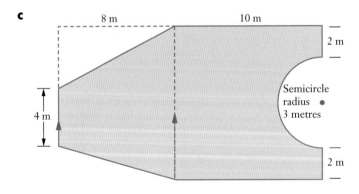

Site prep needed: Yes
Concrete removal needed: Yes
Sealant coating requested: No
Paver type: Limestone

Your turn!

The questions of Exercise 8B have all involved finding the area or perimeter of shapes as applied to situations from real life. Write at least two questions like this of your own invention, researching situations and realistic data from the Internet if necessary, and include answers to the questions.

Inverse questions

The following example, and the exercise that follows, again involve perimeter and area but now you are given one of these quantities and your task is to determine an unknown length. As mentioned in the preface these 'inverse questions' do require some equation solving ability and could be regarded as being beyond the requirements of the syllabus for this unit. I leave it to the reader and to teachers to decide whether to cover them or not.

EXAMPLE 3

If a circle is to have an area of 18 cm^2 what must be the radius of the circle, to the nearest millimetre?

Solution

Let the radius of the circle be r cm.

Thus $\pi r^2 = 18$

and so $r^2 = \dfrac{18}{\pi}$

giving $r = \sqrt{\dfrac{18}{\pi}}$

$= 2.39$ rounded to two decimal places.

The radius of the circle must be 2.4 cm, to the nearest millimetre.

Exercise 8C

1 A square has a perimeter of 64 cm. What is the area of the square?

2 A square has an area of 64 cm^2. What is the perimeter of the square?

3 A rectangle has an area of 36 cm^2. If the rectangle has two sides which are each of length 9 cm what is the perimeter of the rectangle?

4 A square of area 25 m^2 has the same perimeter as a rectangle with a base of length 7 m. Determine the height of the rectangle.

5 If a circle is to have an area of 30 m^2 what should be its radius, to the nearest centimetre?

6 A circle has a circumference of 76 cm. Find the radius of the circle giving your answer to the nearest millimetre.

7 If a circle is to have a circumference of 30 m what should be its diameter, to the nearest centimetre?

8 A square has the same area as that of a circle of radius 18 cm. Find the length of each side of the square giving your answer in millimetres and to the nearest millimetre.

9 Given that each of the following six shapes have the same perimeter, find the values of *a*, *b*, *c*, *d* and *e*. (Assume that angles that appear right angled in the diagrams are indeed right angled.)

Hence write the areas of the shapes in order, from smallest area to largest area.

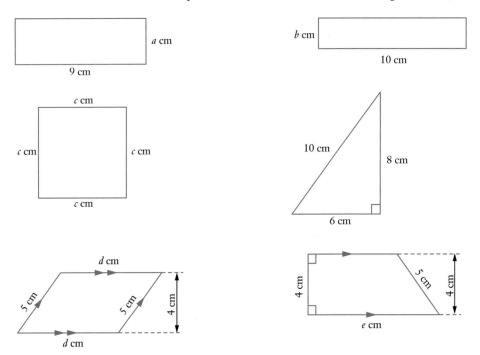

10 Given that each of the following six shapes have the same area, find the values of *a*, *b*, *c*, *d* and *e*. (Assume that angles that appear right angled in the diagrams are indeed right angled.)

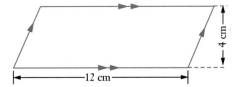

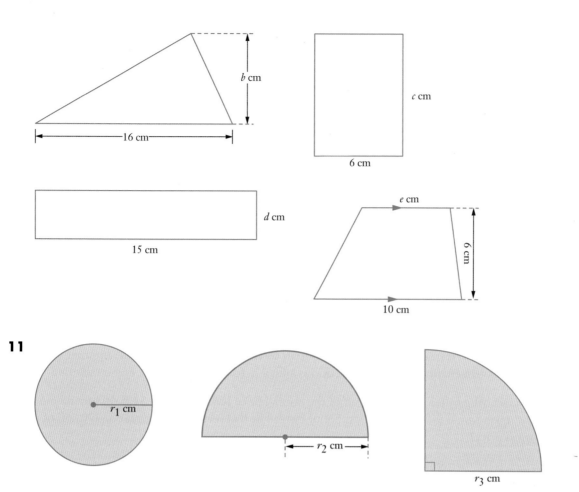

11

Being told that the circle, the semicircle and the quarter-circle shown above all have the same area, Jim initially suggests that $r_3 = 4r_1$ and $r_2 = 2r_1$.

However Jim started to doubt this because, thinking the diagrams might be drawn to scale, clearly $r_3 \neq 4r_1$ and $r_2 \neq 2r_1$.

What is the correct relationship between r_3 and r_1 and between r_2 and r_1?

8. Perimeter and area ●●●●●●●●●●○○

Miscellaneous exercise eight

This miscellaneous exercise may include questions involving the work of this chapter, the work of any previous chapters, and the ideas mentioned in the Preliminary work at the beginning of the book.

1 A Government allowance pays $17 500 per year reducing by 60 cents for each dollar earned over $35 000 per year. What would this allowance pay per year to a person who is eligible for the allowance but who earns

   a $30 000 per year?   **b** $38 000 per year?   **c** $55 000 per year?

2 If $A = \begin{bmatrix} 1 \\ 5 \end{bmatrix}$, $B = \begin{bmatrix} 3 & -2 \end{bmatrix}$ and $C = \begin{bmatrix} 1 & 0 \\ 2 & -1 \end{bmatrix}$ find each of the following matrix products stating clearly if any cannot be determined.

   a CA   **b** BC   **c** AB   **d** BA

3 Using an exchange rate of 1 Australian Dollar (A$) = 0.6572 British Pounds (£) find:

   a how many British Pounds can be bought for A$2000. (Nearest £.)

   b the cost in Australian Dollars when an item costing £560 is ordered on the internet using an Australian credit card, if an extra A$35 is added in fees for the transfer and transaction.

4 After receiving a discount of 15% off of the normal price Jack is charged $105.40 for an item. How much would the item have cost him if instead he had received a discount of 20% off of the normal price?

5 Which is the 'better buy', and why:

   250 g of sliced ham for $4.35   or   450 g of the same sliced ham for $7.95?

6 A right triangle has one side of length 6 cm and another of length 8 cm. What could be the length of the third side?

7 How many times would each of the following wheels rotate in a journey of 10 km?

  a    **b**    **c**

     Diameter 66 cm   Diameter 52 cm   Diameter 74 cm

8 A ladder has a length of 7 metres. How far from the base of a wall of height 6.5 m should the base of the ladder be placed if the top of the ladder is to just reach the top of the wall? (Assume the ground is horizontal and the wall is vertical.)

9 If we neglect air resistance then when something is initially held at rest and released, the distance, d metres, that it has fallen t seconds later is given by:

$$d = 4.9t^2$$

from which it follows that the time taken, t seconds, to fall a distance d metres is given by

$$t = \sqrt{\frac{d}{4.9}}.$$

a A coin dropped from the top of a building reaches the ground 2 seconds later. How high is the building?

b A balloonist inadvertently drops his sandwich from his stationary balloon and sees it hit the ground 6 seconds later. How high was the balloon when the sandwich was dropped (to the nearest ten metres)?

c An elderly person leans out of the window of a high rise building to admire the view and her false teeth fall out! How long will they take to reach the ground if the window was 60 metres above the ground?

10 What percentage of the shape shown on the right is shaded? Give your answer to the nearest 0.1%.

(The shape consists of a circle drawn in a square with the circle just touching all four sides of the square.)

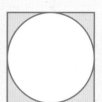

11 A company makes three models, A, B and C, of a particular item. Each model requires a certain number of units of commodities P, Q and R. Matrix X below shows the number of units of each commodity required to make one of each model.

$$
\begin{array}{c}
\\
\text{Model A} \\
\text{Model B} \\
\text{Model C}
\end{array}
\begin{array}{ccc}
\text{P} & \text{Q} & \text{R} \\
\left[\begin{array}{ccc}
2 & 2 & 1 \\
3 & 1 & 1 \\
1 & 3 & 1
\end{array}\right]
\end{array} = X
$$

Each unit of P, Q and R costs the company $50, $60 and $200 respectively.

We could write this as a column matrix, Y:
$$
\left[\begin{array}{c}
50 \\
60 \\
200
\end{array}\right]
$$

or as a row matrix, Z:
$$
\left[\begin{array}{ccc}
50 & 60 & 200
\end{array}\right]
$$

Both XY and ZX could be formed but only one of these will contain information likely to be useful.

a Which is the useful one?

b Form the product.

c Explain the information it displays.

12 Soil conditioner can be added to the soil to improve the nutrient level in the soil and hence aid the growth of anything planted in the soil. A particular brand of soil conditioner is available in 25 kg bags.

The manufacturer advises that the appropriate application rate of this conditioner depends on whether the existing soil requires minimal application, medium application or heavy application.

The advised rates of application for each classification are as follows:

Minimal application: 100 g per square metre
Medium application: 300 g per square metre
Heavy application: 500 g per square metre.

For each of the following areas, how many bags (rounded up to next whole number) would be needed to add the conditioner at the manufacturer's advised application rate with the application classifications as indicated.

a 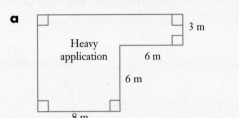   **b** 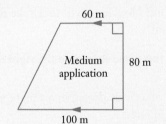   **c**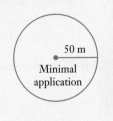

13 The pay slip shown below was created on a spreadsheet. The pay clerk simply enters the bold and italicised entries, i.e. name, normal hourly rate, week and hours and the 'boxed amounts' are automatically completed. Use a spreadsheet to produce and print out this pay slip yourself and then generate similar pay slips for:

Patsy Ling, Week 23, Normal hourly rate $19.20,
Normal hours 35, Time and a half hours 4, Double time hours nil.

Troy Marcesi, Week 23, Normal hourly rate $21.40,
Normal hours 35, Time and a half hours 6, Double time hours 4.

ANGUS SWEENEY					
Normal hourly rate	$17.60	/h	Normal	$17.60	/h
Week	23		Time and a half	$26.40	/h
			Double time	$35.20	/h
Hours worked			Payment due		
Normal	35		$616.00		
Time and a half	3		$79.20		
Double time	3		$105.60		
		Total	$800.80		

Surface area
and volume

Surface area

The *Preliminary work* section at the beginning of this book included a reminder of how to determine the area of rectangles, triangles, parallelograms, trapeziums and circles. Chapter eight then required you to use these skills in various situations.

We can use this ability to determine the area of various shapes to determine the total surface area of prisms, cylinders and pyramids.

For example, consider the solid rectangular prism shown.

Face	Area
Front	20 cm^2
Back	20 cm^2
Top	15 cm^2
Base	15 cm^2
Left side	12 cm^2
Right side	12 cm^2
Total surface area	94 cm^2

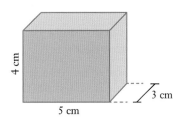

This prism has a total surface area of 94 cm^2.

Prisms

- A prism is a three dimensional shape with all of its faces polygons and with the same shape cross section all along its length.

- Prisms are named according to their uniform cross section. However we do not tend to call a cube a square prism because 'cube' tells us that *all* the faces are square. Not all square prisms are cubes.

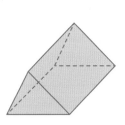
Triangular prism

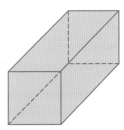
Square prism

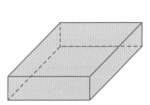
Rectangular prism

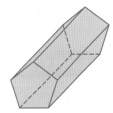
Pentagonal prism

- If all of the sides of a prism go back perpendicularly from the front face all of the sides will be rectangular. Such prisms are said to be *right* prisms. Assume all of the prisms in this unit are right prisms.

- If a three dimensional shape has a circular uniform cross section it is a **cylinder**. Whilst a cylinder is sometimes thought of as a circular prism, technically it is not a prism because its faces are not polygons.

Pyramids

- A three dimensional shape that has a polygon as its base and all the sides meeting at a point is called a pyramid. We name the pyramid according to the shape of its base:

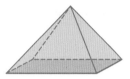
Rectangular pyramid

Pentagonal pyramid

Triangular pyramid

Hexagonal pyramid

- As with prisms we can determine the total surface area of a pyramid by summing the areas of the separate surfaces.

- If a three dimensional shape has a circular base and then comes up to a point it is a **cone**. Whilst a cone is sometimes thought of as a circular based pyramid, technically it is not a pyramid because the base is not a polygon.

For a solid cone with base radius r and slant height l (see diagram), the curved surface area and total surface area are given by the rules:

$$\text{Curved surface area} = \pi r l$$
$$\text{Total surface area} = \pi r l + \pi r^2$$

Spheres

One common three dimensional shape not covered so far in this chapter is the sphere. The surface area of a sphere of radius r is given by:

$$\text{Surface area} = 4\pi r^2$$

EXAMPLE 1

Find the surface area of each of the following solid bodies.

a Triangular prism

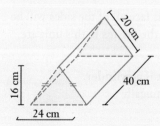

b Trapezoidal prism

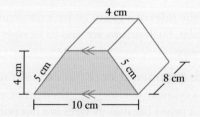

Solution

a Base: 960 cm^2
Ends: 2×192 cm^2
Sides: 2×800 cm^2
Total: 2944 cm^2

The triangular prism has a total surface area of 2944 cm^2.

b Base: 80 cm^2
Top: 32 cm^2
Ends: 2×28 cm^2
Sides: 2×40 cm^2
Total: 248 cm^2

The trapezoidal prism has a total surface area of 248 cm^2.

EXAMPLE 2

Find the surface area of each of the following solid bodies.

a Cylinder

b Hemisphere

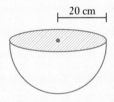

Solution

a 'Unrolling' the cylinder we see that its surface area consists of three parts:

Two circles and a rectangle.

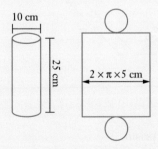

Area of the two circles: $2 \times \pi \times 5^2$ cm^2

Area of the rectangle: $2 \times \pi \times 5 \times 25$ cm^2

Total: 942.5 cm^2
             (to 1 decimal place)

The cylinder has a total surface area of 942.5 cm^2 correct to 1 decimal place.

b Surface area of sphere $= 4\pi r^2$

Thus for the hemisphere:

Area of curved surface: $2 \times \pi \times 20^2$ cm^2

Area of flat surface: $\pi \times 20^2$ cm^2

Total: 3770 cm^2
                   (to the nearest cm^2)

The hemisphere has a total surface area of 3770 cm^2, to the nearest cm^2.

Exercise 9A

Find the surface area of each of the following solids shown in numbers 1 to 16.

1 Cube

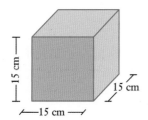

15 cm
15 cm
15 cm

2 Rectangular prism

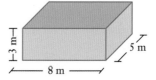

3 m
5 m
8 m

3 Rectangular prism

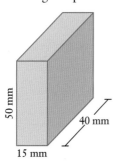

50 mm
40 mm
15 mm

4 Triangular prism

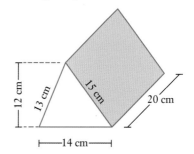

12 cm
13 cm
15 cm
20 cm
14 cm

5 Triangular prism

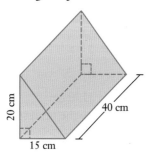

20 cm
40 cm
15 cm

6 Trapezoidal prism

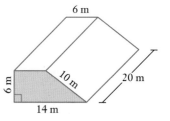

6 m
6 m
10 m
20 m
14 m

7 Hexagonal prism

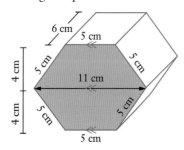

6 cm
5 cm
5 cm
5 cm
4 cm
4 cm
11 cm
5 cm
5 cm
5 cm

8 Pentagonal prism

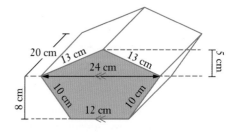

20 cm
13 cm
13 cm
24 cm
5 cm
8 cm
10 cm
10 cm
12 cm

ISBN 9780170390194

9 Square pyramid with top point vertically above centre of base.

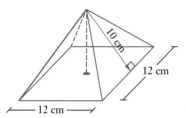

10 Rectangular pyramid with top vertically above centre of base.

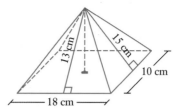

11 Cylinder
Give answer to the nearest 1 cm^2.

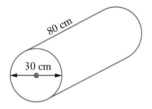

12 Sphere (radius 75 cm)
Give answer to the nearest 100 cm^2.

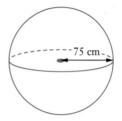

13 Hemisphere.
Give answer to the nearest 50 cm^2.

14 Cone
Give answer to the nearest 100 cm^2.

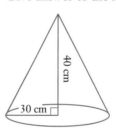

15 Cube on cube

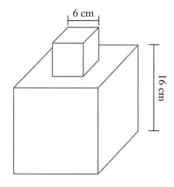

16 Hemisphere on hemisphere
Give answer to the nearest cm^2.

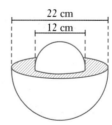

Volume

For a rectangular prism as shown:

$$\text{Volume} = a \text{ cm} \times b \text{ cm} \times \text{length}$$
$$= ab \text{ cm}^2 \times \text{length}$$
$$= \text{area of front face} \times \text{length}$$

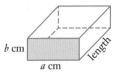

The uniform cross section of triangular prisms, pentagonal prisms and many other prisms are *not* rectangular. However we can find the volume of such prisms in the same way as we can for rectangular prisms, i.e. find the area of the uniform cross section and then multiply by the length of the prism.

Thus for all of the right prisms shown below:

> Volume of prism = Area of uniform cross section × length

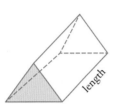 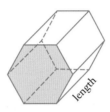 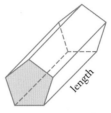 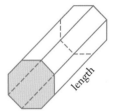

This rule can also be used to determine the volume of other shapes which, because they have some faces that are *not* polygons, are not prisms. The rule can be used provided the shape has a uniform cross section and the 'body' of the shape goes back perpendicularly from the uniform cross section.

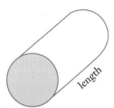 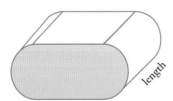  

Thus for all of the above shapes:

> Volume = Area of uniform cross section × length

To determine the volume of a pyramid, we use the rule:

> $$\text{Volume of pyramid (or cone)} = \frac{\text{base area} \times \text{height}}{3}$$

For a sphere, the rule is:

> $$\text{Volume of sphere of radius } r = \frac{4}{3}\pi r^3$$

 ISBN 9780170390194

Find the volume of the rectangular prism shown.

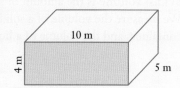

Solution

Volume = shaded area × length
= 4 m × 10 m × length
= 40 m² × 5 m
= 200 m³

The rectangular prism has a volume of 200 m³.

Find the volume of the rectangular based pyramid shown.

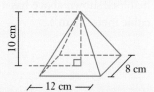

Solution

Volume = $\dfrac{\text{base area} \times \text{height}}{3}$

= $\dfrac{96 \text{ cm}^2 \times 10 \text{ cm}}{3}$

= 320 cm³

The pyramid has a volume of 320 cm³.

96 × 10 ÷ 3	
	320

Find the volume of a sphere of radius 2.41 m giving your answer to the nearest cubic metre.

Solution

Volume of sphere = $\dfrac{4}{3}\pi r^3$

Hence volume = $\dfrac{4}{3} \times \pi \times 2.41^3$ m³

= 58.6 m³ (to 1 decimal place)

The sphere has a volume of 59 m³ (to the nearest cubic metre).

$\dfrac{4}{3} \times \pi \times 2.41^3$	
	58.63267886

9. Surface area and volume ●●●●●●●●●○

Volume and capacity

Whilst **volume** is the amount of space a solid occupies, **capacity** is the amount a container can hold. We measure the volume of a solid in mm³, cm³ and m³ and we tend to measure the capacity of containers and the volume of a liquid in millilitres, litres and kilolitres.

The block shown above has a volume of 1 cubic centimetre (1 cm³).

The container shown above has a **capacity** of **1 millilitre (1 mL)**.

The block shown above has a volume of 1 cubic metre (1 m³).

The container shown above has a **capacity** of **1 kilolitre (1 kL)**.

> 1 L = 1000 mL
> 1 L of liquid occupies 1000 cm³
> 1 kL = 1000 L
> 1 kL of liquid occupies 1 m³

Whilst the units for capacity are based on the litre it is sometimes the case in real life that capacity is given using cm³ or m³. For example the capacity of a motor bike engine may be quoted in 'cc', standing for cubic centimetres. The capacity of a waste removal bin may be quoted in m³.

Exercise 9B

Find the volume of each of the following solids.

1 Cube

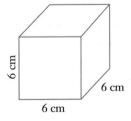

6 cm
6 cm
6 cm

2 Rectangular prism

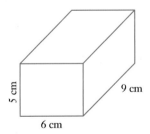

5 cm
9 cm
6 cm

3 Triangular prism

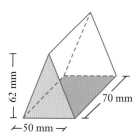

62 mm

70 mm

50 mm

4 Cylinder
Give your answer to the nearest
50 cubic centimetres.

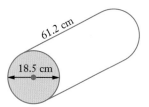

61.2 cm

18.5 cm

5 Rectangular pyramid

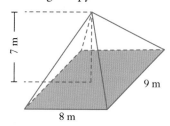

7 m

9 m

8 m

6 Triangular pyramid

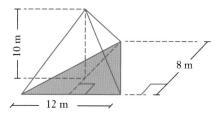

10 m

8 m

12 m

7 Cone
Give your answer to the nearest
cubic centimetre.

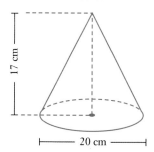

17 cm

20 cm

8 Sphere
Give your answer to the nearest
cubic centimetre.

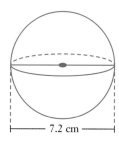

7.2 cm

9 Hemisphere
Give your answer to the nearest
100 cubic centimetres.

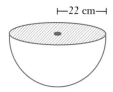

22 cm

10 Triangular pyramid

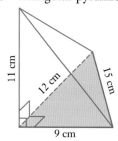

11 cm

12 cm

15 cm

9 cm

Find the capacity of each of the following containers.

11 Rectangular container

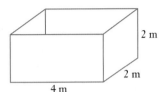

12 Rectangular container

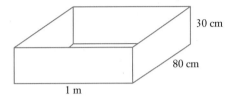

13 Cylindrical container
Give your answer to the nearest millilitre.

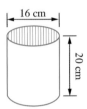

14 Hemispherical container
Give your answer to the nearest litre.

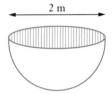

15 A solid metal cube with all edges of length 20 cm is to be melted down and recast into smaller cubes each with an edge length of 1 cm. How many such cubes could be made?
If instead the original cube were to be recast into spheres of radius 1 cm how many such spheres could be made?

16 Find the volume of material required to make the hemispherical shell shown with the radius of the internal hemispherical 'space' and of the external hemisphere as shown.

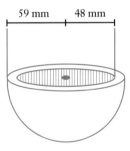

Give your answer rounded **up** to the next 100 mm³.

Applications

In the following exercise you will again need to determine various surface areas, volumes and capacities. However the questions now involve some everyday contexts for which determining these quantities is significant.

Exercise 9C

1 Determine the volume of concrete required to make each of the following concrete slabs.
Give each answer in cubic metres rounded to one decimal place.

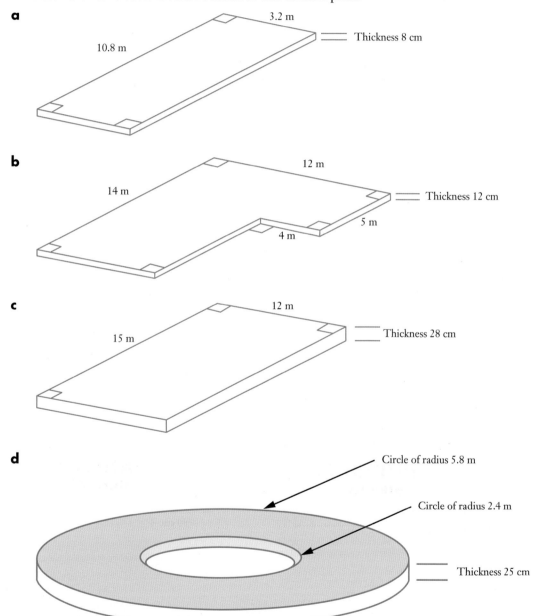

a
3.2 m
10.8 m
Thickness 8 cm

b
12 m
14 m
Thickness 12 cm
5 m
4 m

c
12 m
15 m
Thickness 28 cm

d
Circle of radius 5.8 m
Circle of radius 2.4 m
Thickness 25 cm

2 A skip hire company has skips of various sizes available for hire.
One of the available sizes is as shown.

Confirm that with the dimensions shown the volume of this skip, when filled level to the rectangular top, is 8.91 m^3.

However, because the thickness of the walls makes each dimension of the fillable space somewhat less than the external dimensions given, the company advertises this skip as being an 8 m^3 skip.

For each of the following skips determine the volume of each, when filled level to the rectangular top, according to the dimensions shown, and then suggest what size the company is likely to advertise the skip as being.

a

b

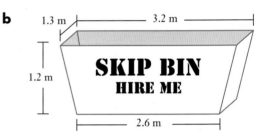

c

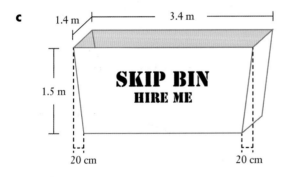

d

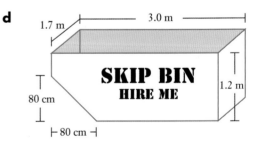

ISBN 9780170390194

3 Find the volume of water needed to fill each of the following swimming pools to a level that is 10 cm below the top of the pool. (Assume that the measurements given in the diagrams are internal dimensions.)

How long would it take to fill each pool using a water supply that delivers 50 litres per minute?

a

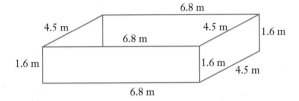

b

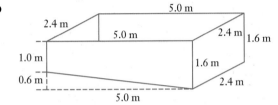

c　The floor of this pool is a rectangle with a semi-circle at each end. The pool is of uniform depth.

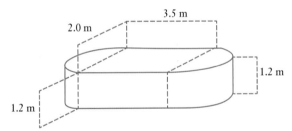

d

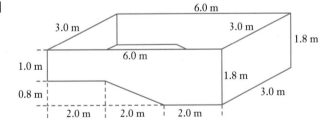

4 A company makes containers of various sizes that can fit onto trucks which are then used for transporting loads of soil or sand or similar material, for example, at mine sites, building sites etc.

Shutterstock.com/bogdanhoda

Neglecting the thickness of the steel used to make the container, determine in cubic metres and rounded to one decimal place, the volume of material each of the following containers can hold for both

   i a load that is level with the top of the container

   ii a heaped load – which increases the amount that can be carried by 25% for container **a**, 30% for containers **b** and **c** and 35% for container **d**.

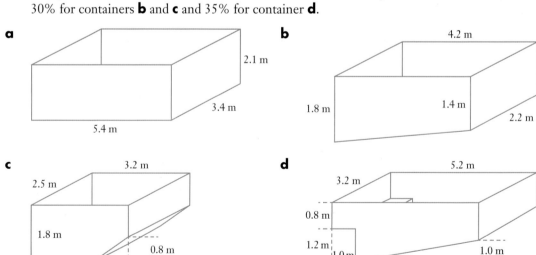

5 a The water tank shown below is to be constructed in concrete, with concrete walls and a concrete base. A metal lid is to be put in place later.

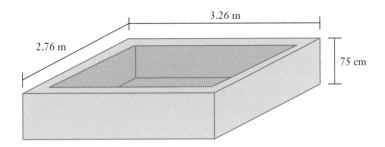

The concrete base is to be 150 mm thick.
The walls are to be 130 mm thick.
Determine the capacity of the water tank and the volume of concrete required to make it.

b The cylindrical water tank shown below is to be constructed in concrete, with the circular base of diameter 5.4 metres. A plastic lid is to be put in place later.

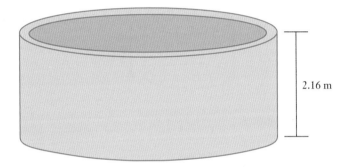

The concrete base is to be 200 mm thick.
The walls are to be 150 mm thick.
Determine the capacity of the water tank and the volume of concrete required to make it.

CSIRO scienceimage/Gregory Heath

6 To calculate the area of metal sheeting required to make cylindrical metal cans a company calculates the external surface area of the cylinder involved and then adds 10% for joins and wastage.

Calculate the area of metal sheeting required for each of the following production runs (to the nearest m² and including base and lid):

a Fifty thousand of the cans shown.

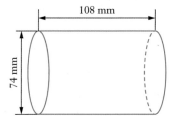

b Two thousand of the cans shown.

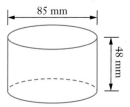

c One million of the cans shown.

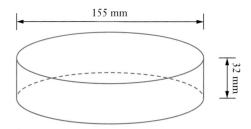

d Five thousand of the cans shown.

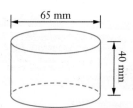

ISBN 9780170390194

7 a Taking the available space inside an oil drum as being cylindrical with a base diameter of 572 millimetres and a height of 851 millimetres determine the capacity of an oil drum in litres, rounded to two decimal places.

b In fact 200 litres of oil is placed into each drum. What percentage of the available space is occupied by oil when a drum contains 200 litres of oil (to the nearest 0.5%)?

c In the United States of America these oil drums are sometimes referred to as 55-gallon drums but in the United Kingdom they are referred to as 44-gallon drums.

Do some research and write a paragraph or two explaining why this is so.

d Some of the big oil tankers that move large quantities of oil around the Earth can hold more than 2 million barrels of oil.

If the oil from 2 million barrels were to form 'a cube of oil' what would be the length of each side of the cube?

8 For each of the following pyramids find

   i the volume of the pyramid,

   ii the weight of rock required to make the pyramid assuming that for the rock used each cubic metre has a weight of 2500 kg and that 10% of the volume of each pyramid is empty space (i.e. chambers and corridors).

a The Pyramid of Khufu.
    Square base of side 230 metres, height 146 metres.

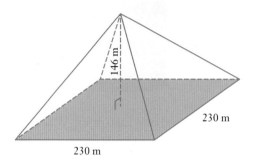

b The Pyramid of Khafra.
    Square base of side 215 metres, height 144 metres.

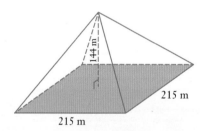

c The Pyramid of Menkaura.
    Square base of side 108 m, height 66 m.

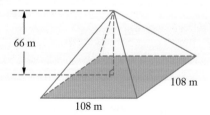

9 Find the volume of each of the following spheres.

a The Earth is approximately a sphere of radius 6400 km.

pixabay.com/skeeze

b A spherical basketball of diameter 24 cm.

Shutterstock.com/tothzoli001

c A spherical wrecking ball of radius 45 cm.

Shutterstock.com/cigdem

d The Moon is approximately a sphere of radius 1740 km.

pixabay.com/Ponciano

e A snooker ball is a sphere of diameter 52.5 mm.

Shutterstock.com/NY Studio

f An eyeball is approximately a sphere of diameter 2.2 cm.

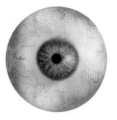

Shutterstock.com/Tischenko Irina

10 Parcel-delivering companies usually charge according to the weight of a parcel. However, very large light parcels can take up a lot of room in the delivery vehicle, denying space for other parcels, whilst not bringing in much revenue. To avoid this problem, when asked to deliver a large light parcel, the cost may be based on the parcel's 'cubic weight'. Each company may have its own formula for calculating their version of the cubic weight. One way involves determining the cubic weight in kilograms by multiplying the parcel's volume in cubic metres by 250. By this method, a rectangular parcel with dimensions 80 cm by 50 cm by 40 cm would have a cubic weight of 40 kg (because $0.8 \times 0.5 \times 0.4 \times 250 = 40$). If the real weight of the parcel was less than 40 kg then it would be charged for as if it weighed 40 kg. If it weighed over 40 kg it would be charged for at its real weight.

Companies may charge for each kg *or part thereof*. A parcel weighing 4.2 kg would be charged as 5 kg. A parcel weighing 17.8 kg would be charged as 18 kg etc.

For each of the following parcels calculate both the real weight and the cubic weight (using the above method), and then calculate the charge.

a Charged at $7.50 for the first kilogram plus $4.40 per kilogram, or part thereof, after that.

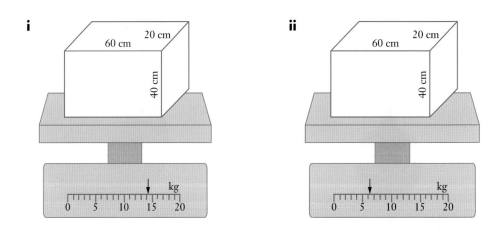

b Charged at $8.40 for the first kilogram plus $5.20 per kilogram, or part thereof, after that.

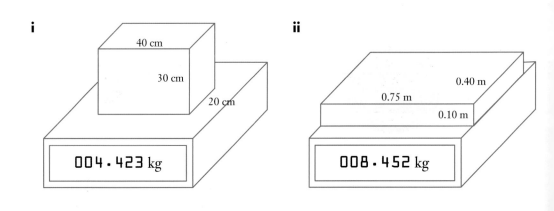

11 A company specialising in pool restoration equipment sells 'epoxy pool coating' that can be painted onto the interior walls and floor of swimming pools.

For each of the pools shown below determine

   i the *total* surface area of the interior walls and floor

   ii the amount of paint required, rounded up to the next whole litre, if three coats are required to all interior surfaces and each 1 litre of the paint will put one coat on an area of 8 square metres.

Note:

- Dimensions shown should be taken as interior dimensions.
- The open top surface of each of the pools shown is horizontal with the first being circular and the other three rectangular.

a

Circle of radius 4 m

1.5 m

b

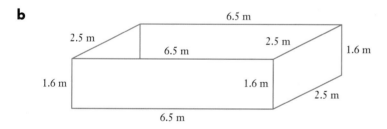

6.5 m

2.5 m

6.5 m

2.5 m

1.6 m

1.6 m

1.6 m

2.5 m

6.5 m

c

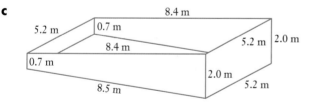

8.4 m

5.2 m

0.7 m

8.4 m

5.2 m

2.0 m

0.7 m

2.0 m

5.2 m

8.5 m

d

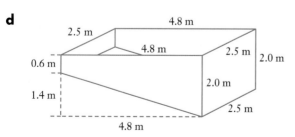

4.8 m

2.5 m

4.8 m

2.5 m

2.0 m

0.6 m

2.0 m

1.4 m

2.5 m

4.8 m

ISBN 9780170390194

12 A company makes rectangular posting boxes using strong, lightweight card. Assuming the thickness of the card is 4 mm find the capacity of each of the following boxes, giving your answers rounded *down* to a whole multiple of five cubic centimetres.

(The diagrams show *external* dimensions.)

(Assume each side, base and top are one thickness of card.)

a

b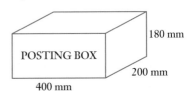

c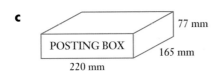

d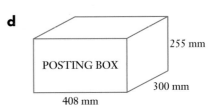

Modelling yourself very simply as two cylindrical arms, two cylindrical legs, a cylindrical body and a spherical head, estimate your surface area.

Compare your answer with that given by Mosteller's formula which claims to give a reasonable estimate of the surface area of a person of weight W kg and height h cm using the rule:

Estimate of surface area in square metres = $\sqrt{\dfrac{W \times h}{3600}}$.

RESEARCH

As mentioned earlier, all of the prisms encountered in this chapter have been assumed to be 'right' prisms. What is an 'oblique' prism and how is its volume determined?

Inverse questions

The following example, and the exercise that follows, again involves surface area or volume but now you are given one of these quantities for an object and your task is to determine an unknown length. (Indeed the last part of question 7 in the previous exercise was a simple one of this type.) As with the similar exercise in the previous chapter these 'inverse questions' do require some equation solving ability and could be regarded as being beyond the requirements of the syllabus for this unit. I leave it to the reader and to teachers to decide whether to cover them or not.

ISBN 9780170390194

EXAMPLE 6

A solid cube has a total surface area of 384 cm^2.
Determine the length of each edge of the cube.

Solution

Let each edge of the cube be x cm.

Each face will then have an area of x^2 cm^2.

Hence $\qquad 6x^2 = 384$.
Divide by 6: $\qquad x^2 = 64$.

Solving and rejecting the negative solution gives $x = 8$.

Each edge of the cube is of length 8 cm.

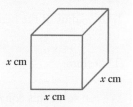

EXAMPLE 7

A solid cylinder with circular ends of radius 30 mm
has a total surface area of 52 800 mm^2, to the nearest 100 mm^2.

Find the length of the cylinder giving your answer to the
nearest millimetre.

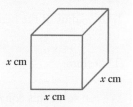

Solution

Let the length of the cylinder be x mm, as shown on the right.

Area of the two circles: $2 \times \pi \times 30^2$ mm^2

Area of curved surface: $2 \times \pi \times 30 \times x$ mm^2

Hence: $\qquad 2 \times \pi \times 30^2 + 2 \times \pi \times 30 \times x = 52\,800$

Take 1800π from each side: $\qquad 60\pi x = 52\,800 - 1800\pi$

Divide each side by 60π: $\qquad x = \dfrac{52\,800 - 1\,800\pi}{60\pi}$

$\qquad\qquad\qquad\qquad = 250$ mm (to the nearest millimetre).

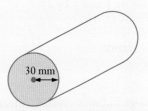

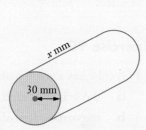

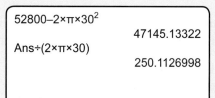

solve(2·π·30^2+60·π·x = 52800, x)
$\qquad$ {x = 250.1126998}

52800−2×π×30^2
$\qquad\qquad\qquad$ 47145.13322
Ans÷(2×π×30)
$\qquad\qquad\qquad$ 250.1126998

EXAMPLE 8

A hemispherical bowl can hold 0.5 litres of liquid when filled level to the brim. Find the radius of the hemisphere.

Solution

If the radius of the hemisphere is r cm then volume $= \frac{2}{3}\pi r^3$ cm^3

Therefore: $\quad\quad\quad\quad\quad\quad\quad \frac{2}{3}\pi r^3 = 500$

We can solve this equation by using the solve facility of some calculators as shown on the right.

It can also be solved by algebraic manipulation.

$$\frac{2}{3}\pi r^3 = 500$$

$\times$ by 3 and $\div$ by 2π: $\quad\quad\quad r^3 = \frac{500 \times 3}{2\pi}$

Hence: $\quad\quad\quad\quad\quad\quad\quad r = \sqrt[3]{\frac{500 \times 3}{2\pi}}$

$$\approx 6.204$$

solve $\left(\frac{2}{3}\cdot\pi\cdot x^3 = 500,\ x \right)$

$\{x = 6.203504909\}$

The radius of the hemisphere is 62 mm, to the nearest millimetre.

Exercise 9D

1 A cube has a volume of 125 cm^3. Find:
  a the length of each side of the cube
  b the surface area of the cube.

2 A cube has a surface area of 96 cm^2. Find:
  a the length of each side of the cube
  b the volume of the cube.

3 Find, to the nearest millimetre, the radius of a sphere given that it has a volume of 74 cm^3.

4 A hemispherical bowl can hold 1.25 litres of liquid when filled level to the brim. Find the radius of the hemisphere giving your answer to the nearest millimetre.

5 A solid sphere has a surface area of 7660 cm^2. Find the radius of the sphere giving your answer to the nearest millimetre.

6 A cube has a volume of 3375 cm^3. Find the surface area of the cube.

7 A sphere has a volume of 3375 cm^3. Find the surface area of the sphere.

8 Each of the following solids have the same volume. Find the values of v, w, x, y and z.
(For y and z give your answer correct to one decimal place.)

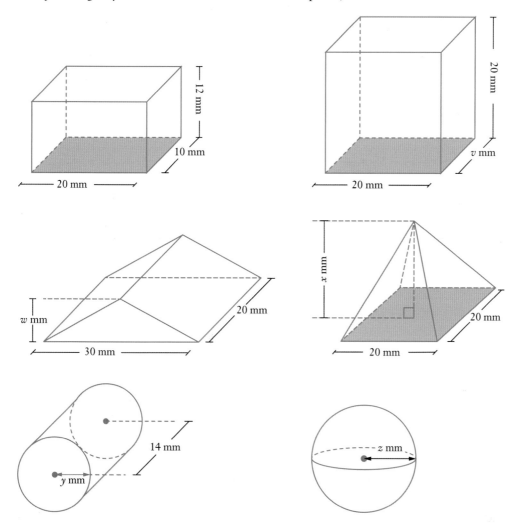

9 A solid cylinder with circular ends each of radius 74 mm has a total surface area of 133 900 mm², to the nearest square millimetre.

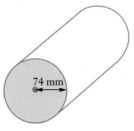

Find the length of the cylinder giving your answer to the nearest millimetre.

10 Each of the two solids shown have the same surface area as a solid cube of side 10 cm.

Find the values of x and y.

a Square pyramid with a top point vertically above the centre of the base.

b Trapezoidal prism

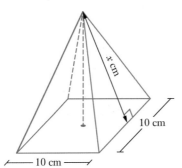

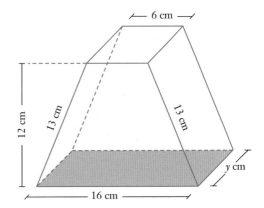

11 Given that the solid rectangular prism shown has a total surface area of 137.5 cm², find the value of x.

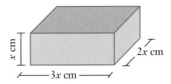

12 A solid metal sphere of radius 20 cm is recast into two identical smaller spheres. To the nearest millimetre, and assuming no loss of metal in the process, what will be the radius of each of these two spheres?

How does the total surface area of the two smaller spheres compare with the surface area of the original sphere?

13 A solid metal sphere of radius 21.0 cm is to be melted down and recast into two cylinders, each with end radius of 14.0 cm, but with one cylinder twice the length of the other. Assuming no loss of metal in the process, what will be the length of each cylinder?

How does the total surface area of the two cylinders compare with the surface area of the original sphere?

14 A container is in the shape of an upturned cone.

As liquid is poured into this container, the ratio of the radius of the curved surface of the liquid to the depth of the liquid is always 1 : 2. That is, for the situation shown $r : d = 1 : 2$.

What is the depth of the liquid (d in the diagram) when the container holds 0.5 litres of liquid?

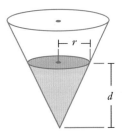

Miscellaneous exercise nine

This miscellaneous exercise may include questions involving the work of this chapter, the work of any previous chapters, and the ideas mentioned in the Preliminary work at the beginning of the book.

1 Each part of this question involves a student working out an answer which is marked and commented on by their teacher. For each situation write some explanation to the student of what it is they are doing wrong, what they should do to get the answer right and include some examples that show correct responses.

a Jamie is asked to work out $856\,000 \times 5\,200\,000$.
Jamie uses his calculator and his written answer (in blue) and his teacher's comment (in red) are shown below:

> $856\,000 \times 5\,200\,000 = 4.512\text{E}+12$ ✗
> We do not write our answer in this way, Jamie.

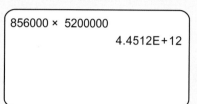

856000 × 5200000

4.4512E+12

b Shania is asked to find the decimal equivalent of five sevenths.
Shania uses her calculator and her written answer and her teacher's comment are shown below:

> $5 \div 7 = 0.714\,285\,7143$ ✗
> This is an approximation of the decimal equivalent, Shania.

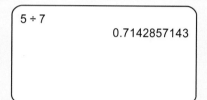

5 ÷ 7

0.7142857143

c Jin is asked to work out, to the nearest dollar, the total cost of 15 identical items if 28 of these items cost a total of $623. Jin's written answer and the teacher's comment are shown below:

> $\$623 \div 28 = \$22$ to the nearest $\$1$.
> $15 \times \$22 = \$330$.
> Fifteen of the items would cost $\$330$, to the nearest dollar. ✗
> Avoid premature rounding, Jin.

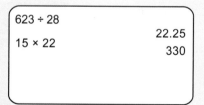

623 ÷ 28

15 × 22

22.25

330

d Junior is asked: If a 433 cm length of wood is to be hand sawn into 7 pieces of equal length how long will each piece be?
Junior's answer and his teacher's comment are shown below:

> Each piece will be $61.857\,142\,86$ cm long. ✗
> Inappropriate accuracy claimed, Junior.

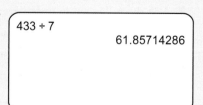

433 ÷ 7

61.85714286

2 After 10% Goods and Sevices Tax (GST) has been added the total cost of an item is $129.14. What was the cost of the item before GST was added?

3 Anne's row of the matrix shown indicates that she is qualified to nurse in *Accident & Emergency* (A&E), and in *Intensive Care* (as indicated by the 1s) but not in *Midwifery* (indicated by 0).

The lines from the name Jo in the diagram indicate that Jo is qualified to work in *A&E* and *Midwifery* but not *Intensive Care*.

Complete both the matrix and the diagram so that they are consistent with each other.

	A & E	Mid	Int/Care
Anne	1	0	1
Jo			
Robyn	0	0	1
Rosemary			

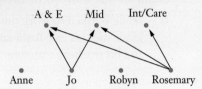

4 Evaluating $2 + 3 \times 5^2$ gives an answer of 77. Insert brackets to make each of the following statements true:

   a $2 + 3 \times 5^2 = 227$ **b** $2 + 3 \times 5^2 = 289$ **c** $2 + 3 \times 5^2 = 125$

5 If $A = \begin{bmatrix} 5 & 2 \\ 1 & 0 \end{bmatrix}$, $B = \begin{bmatrix} 2 & -1 \end{bmatrix}$ and $C = \begin{bmatrix} 1 & 1 \end{bmatrix}$, determine $BA + 3C$.

6 Determine the best buy for the breakfast cereal deals shown.
(Each deal involves the same type of cereal.)

          $11.30 $5.90 $3.70
          825 g   450 g   250 g

7 List the following in order, greatest interest earned first, and state the interest earned in each case.
- $5000 invested for 5 years at 8.5% per annum simple interest.
- $5000 invested for 5 years at 8% per annum compounded every 3 months.
- $5000 invested for 5 years at 8.2% per annum compounded every 6 months.

8 a Assuming the Earth to be a sphere of radius 6.37×10^3 km, find the surface area of the Earth.

   b Approximately 70% of the Earth's surface is covered by water. How many square kilometres of the Earth's surface is covered by water?

9 Copy and complete the following table:

	Unit cost	Number ordered	Sub total	Less 20% discount	Plus 10% tax
e.g.	$34.50	17	$586.50	$469.20	$516.12
	$8.50	23			
	$145.50		$1891.50		
		56		$358.40	
		8			$7208.96

10 a What percentage profit is made when an item that cost $247 is sold for $328? (Give your answer to the nearest percent.)

b What must an item that cost $24 be sold for to make a profit of at least 30%?

c An item sold for $2842 gave the seller a profit of 16% on what he purchased the item for. How much did the seller purchase it for?

d An item sells at auction for $4480. After the auctioneer's commission of 12% is taken from this amount, what remains gives the person who put the item into the auction a profit of 28% on what they purchased the item for. How much did the person putting the item into the auction pay for the item originally?

11 If $A = \begin{bmatrix} -2 & -1 & 2 \\ 1 & -1 & 3 \\ 3 & 2 & -2 \end{bmatrix}$ and $B = \begin{bmatrix} -4 & 2 & -1 \\ 11 & -2 & 8 \\ 5 & 1 & 3 \end{bmatrix}$, determine AB, without the assistance of your calculator.

12 A square piece of glass has a diagonal of length 150 cm. Find the area of the square.

13 A rectangular picture frame is twice as long as it is wide and its diagonals are each of length 30 cm. Find the area of the rectangle.

14 What percentage of the circle shown is shaded? Give your answer to the nearest 1%.

(The shape consists of a square drawn in a circle with all four vertices of the square just touching the circumference of the circle.)

15 The rectangular prism shown is to be cut into two equal pieces.

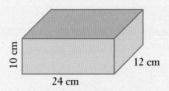

Find the new total surface area, as a percentage of the total surface area of the original prism (to the nearest percent) in each of the following situations:

a

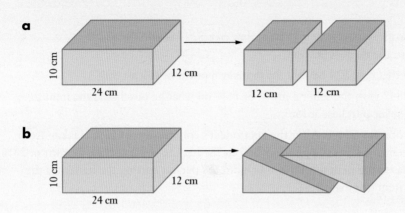

b

16 Earlier in this book it was stated that the curved surface area of a cone of base radius r and slant height l is given by the rule:

$$\text{Curved surface area} = \pi r l.$$

By considering this curved surface to be formed from a sector of a circle of radius l, as shown, try to prove the above rule correct.

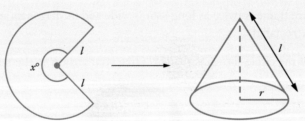

ISBN 9780170390194

10.

Similarity

- Similarity
- Drawing enlargements (and reductions)
- Conditions for similarity
- Similar triangles
- Miscellaneous exercise ten

For the rectangular prism shown:

- The dimensions are 2 cm by 5 cm by 6 cm
- The surface area $=$ 2×5 cm $\times 6$ cm $+$ 2×2 cm $\times 5$ cm $+$ 2×2 cm $\times 6$ cm

                $=$ 60 cm^2 $+$ 20 cm^2 $+$ 24 cm^2

                $=$ 104 cm^2

- The volume $=$ 2 cm $\times 5$ cm $\times 6$ cm

                $=$ 60 cm^3

Now suppose we make a new prism that is twice as long as the original:

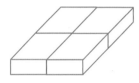

as well as being twice as wide as the original:

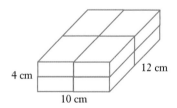

as well as being twice as high as the original:

- The dimensions are now 4 cm by 10 cm by 12 cm
- The surface area $=$ 2×10 cm $\times 12$ cm $+ 2 \times 4$ cm $\times 10$ cm $+ 2 \times 4$ cm $\times 12$ cm

                $=$ 416 cm^2

- The volume $=$ 4 cm $\times 10$ cm $\times 12$ cm

                $=$ 480 cm^3

Thus while   original lengths   :   final lengths $= \quad 1:2,$

the ratio   original surface area   :   final surface area $= \quad 1:4$ i.e. $1:2^2$

and   original volume   :   final volume $= \quad 1:8$ i.e. $1:2^3$

If one three dimensional object is an enlargement of another such that the ratio

lengths in original object : corresponding lengths in enlargement $= 1 : k$

then   surface area of original : surface area of enlargement $= 1 : k^2$

and   volume of original : volume of enlargement $= 1 : k^3$

The quantity k is called the **scale factor** of the enlargement.

For example, suppose we have an oil tanker that is capable of carrying 250 000 barrels of oil.

250 000 barrels

If we made a bigger version of this tanker with all lengths 2 times as long as they are in the original (i.e. the scale factor is 2, and the new tanker will be twice as long, twice as wide and twice as high as the original) the volume of the bigger tanker will be 8 ($= 2^3$) times the volume of the original.

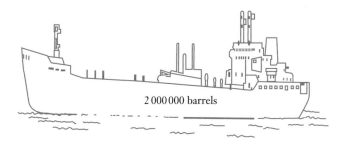

2 000 000 barrels

And the area of sheet metal required to make this bigger tanker will be 4 ($= 2^2$) times the area of sheet metal required to make the original.

If we could make a tanker with all lengths 3 times the original lengths (scale factor of 3) then the volume of this bigger tanker will be 27 ($= 3^3$) times the volume of the original!

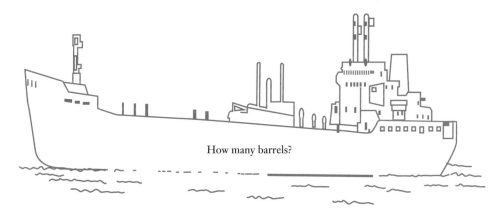

How many barrels?

Exercise 10A

1 The cost of the concrete required to make the slab shown below left is $120.

What does this suggest the cost of the concrete required to make the slab shown below right would be?

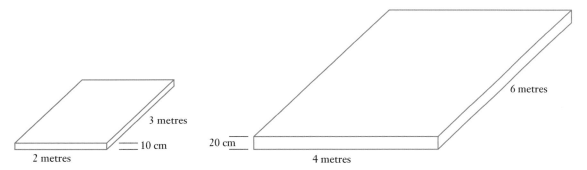

2 The cost of a particular type of paint required to paint the outside of the solid cube shown below left is $30. What does this suggest would be the cost of the same type of paint used to paint the solid cube shown below right?

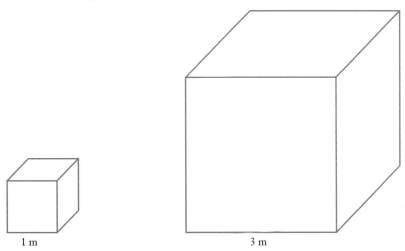

3 It required 2 hours to fill the hemisphere shown below left with water.

What does this suggest for the time the same supply would take to fill the hemisphere below right?

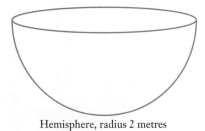

Hemisphere, radius 2 metres

Hemisphere, radius 1 metre

10. Similarity ●●●●●●●●●

4 The cost of electro-plating all surfaces of the solid rectangular girder shown below left was $120. What does this suggest would be the price of electro-plating the girder shown below right with the same material?

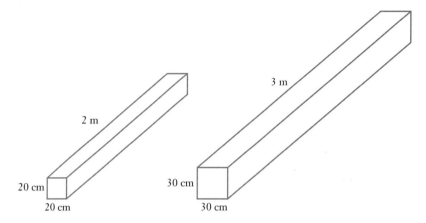

2 m

20 cm

20 cm

3 m

30 cm

30 cm

5 The cost of the card used to make 1000 boxes, all of a particular size, was $28. What does this suggest would be the cost of the same thickness and type of card used to make 5000 boxes each with side lengths twice the length of the corresponding sides of the first mentioned box?

6 A map has the scale 1 : 50 000 which means 1 unit of length on the map is equivalent to 50 000 of those units of length in real life. For example a distance of 1 cm on the map represents 50 000 cm (= 500 metres) in real life.

An area of parkland occupies 4.5 cm^2 on the map. What is the area of this parkland in real life, in square kilometres?

7 A solid sphere of radius 1 metre is made of a particular metal. The metal contained in the sphere is worth $70 000. What does this suggest the same metal forming a solid sphere of radius 2 metres would be worth?

8 A company making mining shovels, a machine used for digging and loading earth or rock on a mine site, makes two types, *The Little Joe* and *The Big John*. Each Big John is simply a scaled up version of a Little Joe with all lengths on the Big John being 1.2 times the equivalent length on a Little Joe. The bucket on a Little Joe has a capacity of 50 m^3. What is the capacity of the bucket on a Big John?

9 A map has a scale of 1 : 10 000, i.e. 1 cm on the map represents 10 000 cm on the ground. What area on the map will a real life area of 8000 m^2 occupy?

10 A model of a sailing boat is made to the scale 1 : 50, i.e a length of 1 cm on the model represents 50 centimetres on the real yacht. If one of the sails on the real yacht has an area of 5 m^2 what will be the area of the same sail on the model?

ISBN 9780170390194

11 An object B is an enlargement of object A such that

$$\text{Volume of object B : Volume of object A} = 27 : 8.$$

What would the following ratios be?

a Length measured on object B : Corresponding length measured on object A.

b Surface area of object B : Surface area of object A.

12 An object C is an enlargement of object D such that

$$\text{Surface area of object C : Surface area of object D} = 16 : 25.$$

What would the following ratios be?

a Length measured on object C : Corresponding length measured on object D.

b Volume of object C : Volume of object D.

Similarity

If one 3-dimensional object is simply a scaled enlargement, or reduction, of another 3-dimensional object then the two objects are said to be **similar**. The ratio of any length on one of the objects to that of the corresponding length on the other object will be the same for all such pairs of lengths. Furthermore, any angle measured on one of the two objects will be *the same size* as the equivalent angle measure on the enlarged (or reduced) object.

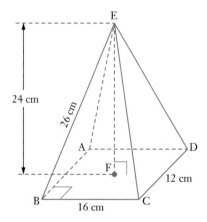
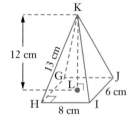

Corresponding lengths in same ratio

$$BE : HK = 2 : 1$$
$$BC : HI = 2 : 1$$
$$CD : IJ = 2 : 1$$
Etc

Corresponding angles equal

$$\angle ABC = \angle GHI \quad (= 90°)$$
$$\angle EFB = \angle KLH \quad (= 90°)$$
$$\angle EBC = \angle KHI$$
Etc

Note also that

	volume of pyramid ABCDE	:	volume of pyramid GHIJK
$=$	$\dfrac{16\text{ cm} \times 12\text{ cm} \times 24\text{ cm}}{3}$	:	$\dfrac{8\text{ cm} \times 6\text{ cm} \times 12\text{ cm}}{3}$
$=$	1536 cm^3	:	192 cm^3
$=$	8	:	1
	i.e. 2^3	:	1

as we would expect from the earlier work of this chapter.

In the same way, two 2-dimensional shapes for which one is a scaled enlargement or reduction of the other are said to be **similar**. Again corresponding sides will be in the same ratio and corresponding angles will be equal.

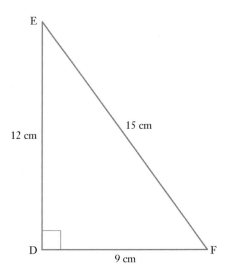

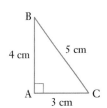

Corresponding lengths in same ratio

AB : DE = 1 : 3

AC : DF = 1 : 3

BC : EF = 1 : 3

Corresponding angles equal

∠BAC = ∠EDF   (= 90°)

∠ABC = ∠DEF

∠BCA = ∠EFD

Triangle ABC is similar to triangle DEF.

We write: △ABC ~ △DEF

The order of the letters **is** important. It would not be correct to say that △ABC ~ △EFD for example. The corresponding vertices, A and D, B and E, C and F, should appear in corresponding places in the similarity statement.

Note also that for the two triangles shown above, that whilst

Length in △ABC : Corresponding lengths in △DEF	=	1 : 3,
Area of △ABC : Area of △ABC	=	$6 \text{ cm}^2 : 54 \text{ cm}^2$
	=	1 : 9
	=	$1 : 3^2$

as we would expect from earlier work in this chapter.

Note: An enlargement may also be referred to as a **dilation**. The medical world, for example, uses the terms   vasodilation, pupillary dilation and   cervical dilation.

Exercise 10B

1 Given that triangles PQR and ZYX shown below are similar, determine

   a the length of RQ,

   b the size of ∠ZXY,

   c the ratio area of triangle PQR : area of triangle ZYX.

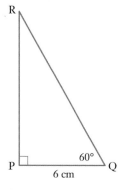 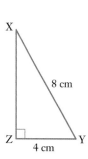

2 Given that in the diagram below △ABC ~ △DEF, determine

   a the length of DE,

   b the size of ∠ACB,

   c the ratio area of triangle ABC : area of triangle DEF.

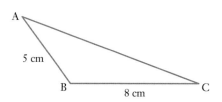 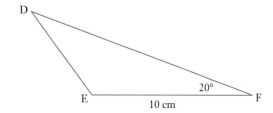

3 Given that the two square based pyramids shown below are similar, determine

   a the length of KL,

   b the ratio surface area of pyramid ABCDE : surface area of pyramid GHIJK,

   c the ratio volume of pyramid ABCDE : volume of pyramid GHIJK.

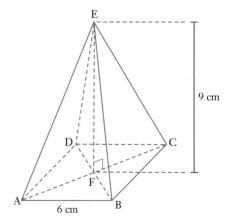 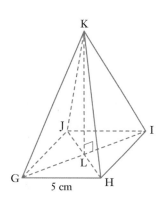

4 Measure distances on the map of Australia shown below to obtain approximate answers for the straight line distances from

a Perth to Sydney,

b Perth to Darwin,

c Melbourne to Sydney,

d Adelaide to Darwin,

e Canberra to Brisbane.

f On the map Sydney is approximately due east of Adelaide. Will this also be the case in real life?

ISBN 9780170390194

5 The diagram below shows the floor plan of a house.

Interpreting an office plan

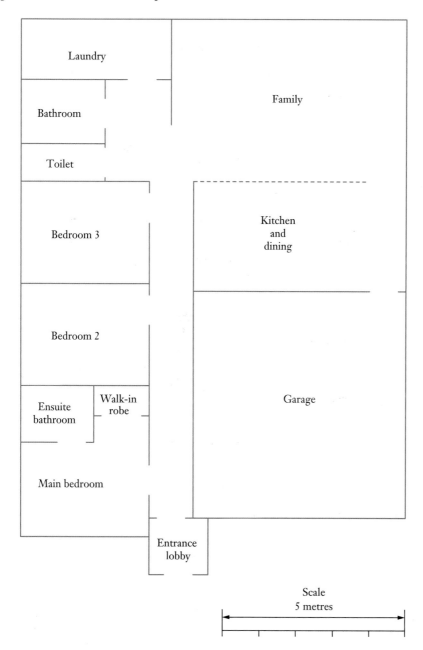

Measure appropriate distances to determine the dimensions of

a the garage,

b the main bedroom (including ensuite and walk-in robe),

c bedroom 3,

d the laundry.

6 The map shown below shows the layout of the Mathematics and Science section of a University Campus.

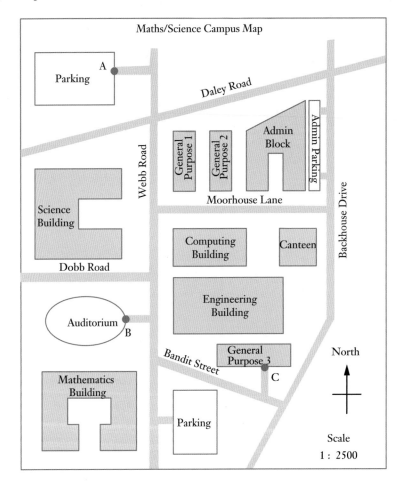

Maths/Science Campus Map

Make appropriate measurements to determine each of the following.

a How far is it in real life from the point marked A, situated at the exit to the Northern car park, to the point marked B, the entrance to the Auditorium, travelling along the road sections by the most obvious route?

b How far is it in real life from the point marked A, situated at the exit to the Northern carpark, to the point marked C, the entrance to building 'General Purpose 3', travelling the road route that involves Webb Road and Bandit Street.

c What is the area in real life of the 'footprint' of the Computing building?

d What is the area in real life of the 'footprint' of the Science building?

ISBN 9780170390194

Drawing enlargements (and reductions)

Nowadays, with so many of the two dimensional diagrams we create being drawn on a computer it can be an easy task to make an enlargement or reduction of the drawing by 'dragging' one corner of the diagram appropriately. (However, if we do want an enlargement of a diagram we must be careful that this 'dragging' doesn't just stretch the diagram either horizontally of vertically, producing a 'stretch' rather than a true enlargement.)

Alternatively a scaled drawing could be constructed using grid paper.

If we have a picture drawn on grid paper, as shown on the right, then by carefully redrawing what is in each square, but on a larger grid, an enlargement can be produced, as shown below.

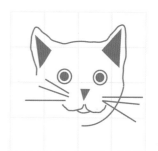

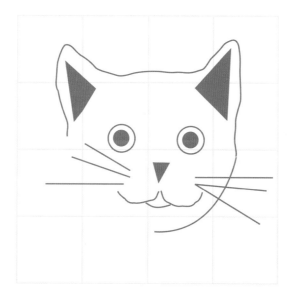

Conditions for similarity

When we create an enlargement or reduction of a shape the enlargement or reduction is *similar* to the original because:

- all corresponding lengths are in the same ratio

and • all corresponding angles are equal.

If instead we are given two shapes and are asked if they are similar we need to check that corresponding sides are in the same ratio and corresponding angles are equal. If both of these conditions are met then the two shapes are similar.

Notice that in general we *cannot* assume similarity if just one of the conditions

corresponding sides in the same ratio

corresponding angles are equal

hold. We need both of these conditions to be the case for similarity.

To illustrate this consider the following:

The two rectangles shown below may have their angles matching but the rectangles are not similar.

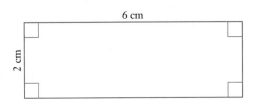

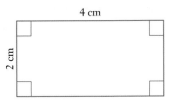

The parallelograms shown below may have sides in the same ratio but the parallelograms are not similar.

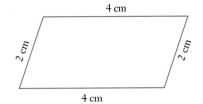

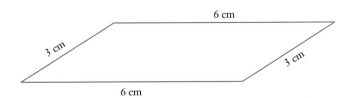

Note: The statement at the top of this page did say 'in general'. For triangles the situation is a little different, as we shall see later in this chapter.

Exercise 10C

For each of questions 1 to 8, state whether or not the two shapes shown are *similar* or *not similar* and for those that are state the ratio

lengths in shape A : corresponding lengths in shape B.

1

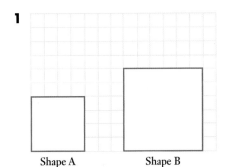

Shape A   Shape B

2

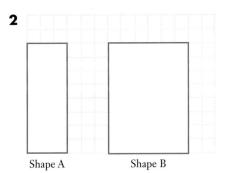

Shape A   Shape B

3

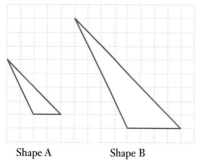

Shape A   Shape B

4

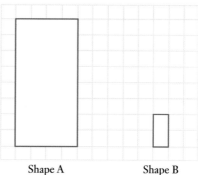

Shape A   Shape B

5

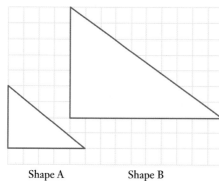

Shape A   Shape B

6

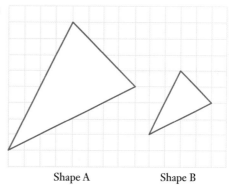

Shape A   Shape B

7

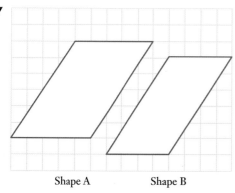

Shape A   Shape B

8

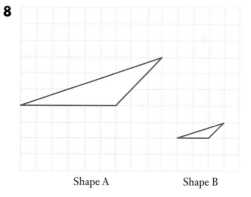

Shape A   Shape B

9 Are all triangles similar to each other?

10 Are all rectangles similar to each other?

11 Are all squares similar to each other?

12 Are all isosceles triangles similar to each other?

13 Are all right-angled triangles similar to each other?

14 Are all equilateral triangles similar to each other?

Finding sides in
similar triangles

Similar triangles

As was mentioned earlier, triangles do *not* fit the more general condition that to know that two shapes are similar we must check *both* that corresponding sides are in the same ratio *and* that corresponding angles are equal. Indeed to know whether two triangles are similar we can:

- See if the three angles of one triangle are equal to the three angles of the other triangle.

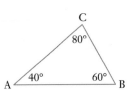

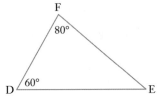

Noticing that each triangle has angles of 40°, 60° and 80° we can say that △ABC and △EDF are similar. (Note carefully the order of the letters. The second triangle is listed as 'EDF' to match the corresponding angles in 'ABC'.)

We write: △ABC ~ △EDF   Reason: Corresponding angles equal.

OR:

- See if the lengths of corresponding sides are in the same ratio.

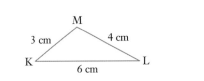

KL : PQ = 6 : 9   LM : QR = 4 : 6   KM : PR = 3 : 4.5
      = 2 : 3   = 2 : 3   = 2 : 3

Thus △KLM and △PQR are similar.

We write: △KLM ~ △PQR   Reason: Corresponding sides in same ratio.

OR:

- See if the lengths of two pairs of corresponding sides are in the same ratio and the angles between the sides are equal.

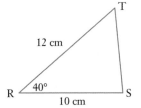

XZ : RS = 5 : 10   XY : RT = 6 : 12
      = 1 : 2   = 1 : 2

The angle between XY and XZ = the angle between RT and RS.

Hence △XZY ~ △RST   Reason: Two pairs of corresponding sides in same ratio and the *included* angles equal.

(198) **MATHEMATICS APPLICATIONS** Unit 1   ISBN 9780170390194

1 The fact that the angles of a triangle have a sum of 180° means that once we have shown that two angles of one triangle are equal in size to two angles in another triangle the third angles must be equal. The condition that corresponding angles are equal is then satisfied and the triangles are similar.

2 If the two triangles are right angled the corresponding sides that are in the same ratio need not *include* the right angle.

EXAMPLE 1

Explain why the two triangles shown below are similar and hence find x.

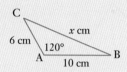 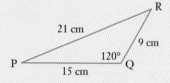

Solution

$$AB : QP = 10 : 15 \qquad\qquad AC : QR = 6 : 9$$
$$= 2 : 3 \qquad\qquad\qquad\quad = 2 : 3$$
$$\angle CAB = \angle RQP \ (= 120°)$$

Thus $\qquad \triangle ABC \sim \triangle QPR \qquad$ Reason: Two pairs of corresponding sides in same ratio and the included angles equal.

Hence: $\quad CB : RP = 2 : 3$
$$x : 21 = 2 : 3$$
but $\qquad 14 : 21 = 2 : 3$
Hence $\qquad x = 14$

EXAMPLE 2

In the diagram, DE = 3 m, EF = 6 m, and EH = 2 m. Find the length of FG, justifying your answer.

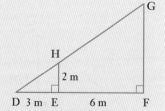

Solution

In triangles DEH and DFG: $\quad \angle HDE = \angle GDF \quad$ (same angle)
$$\angle HED = \angle GFD \quad (= 90°)$$

Hence the third angles will be equal and so $\triangle DEH \sim \triangle DFG$, corresponding angles equal.

Hence $DE : DF = EH : FG$

Letting the length of FG be x cm:
$$3 : 9 = 2 : x$$
i.e. $\qquad 1 : 3 = 2 : x$
But $\qquad 1 : 3 = 2 : 6$
Hence $\qquad x = 6$

FG is of length 6 cm.

Exercise 10D

For questions 1 to 8 of this exercise state whether the two triangles are similar or not and, **for those that are**, name the similar triangles, explain why they are similar, state the ratio of their areas and determine the values of x and y as appropriate.

> **Note**
>
> Within each question any angles marked • are the same size as each other,
>
> any angles marked •• are the same size as each other,
>
> any angles marked ••• are the same size as each other.

1

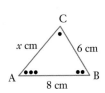

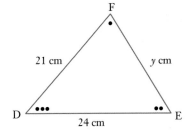

2

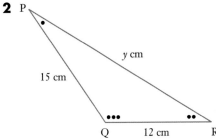

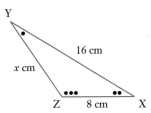

3

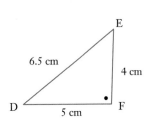

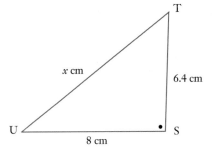

4

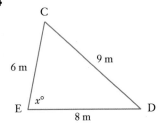

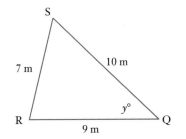

 ISBN 9780170390194

5

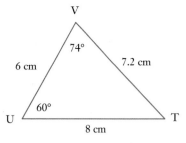

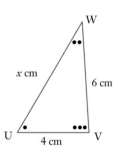

6

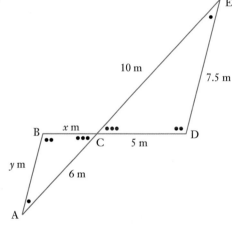

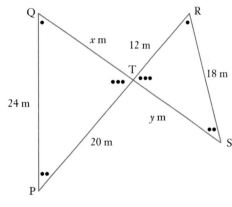

7

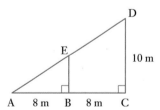

8

9 Determine the length of EB given the information in the diagram shown on the right. (Justify your answer.)

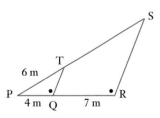

10 Determine the length of TS given the information in the diagram shown on the right. (Justify your answer.)

Miscellaneous exercise ten

This miscellaneous exercise may include questions involving the work of this chapter, the work of any previous chapters, and the ideas mentioned in the Preliminary work at the beginning of the book.

1 When evaluating πr^2 we multiply a number, π, by a length, r, and then by a length, r, again. Multiplying two lengths together give units of area. Multiplying by π, which is just a number and has no units, will not change the units of the answer. Thus even if we did not recognise πr^2 as the formula for the area of a circle of radius r we could still identify it as a formula that will give an answer involving units of area.

In each of the following a, b, c, h and r all represent lengths and any numbers and π are 'unitless'. For each part, state whether the answer would be a length, an area or a volume.

a	$a + b$	**b**	$a + b + c$	**c**	c^2
d	ab	**e**	bc	**f**	$2a + b$
g	abc	**h**	$\dfrac{bh}{2}$	**i**	$3c^2$
j	$\dfrac{(a+b)}{2}h$	**k**	$4\pi r^2$	**l**	$2\pi r$
m	$\dfrac{\pi r^2 h}{3}$	**n**	$\dfrac{4}{3}\pi r^3$	**o**	$\dfrac{1}{2}abh$

2 Express each of the following numbers in scientific notation (i.e. express each in the form $A \times 10^n$ where A is a number between 1 and 10 and n is an integer).

a	1 230 000	**b**	0.0012
c	25 000	**d**	0.000 000 024 5
e	15 000 000 000	**f**	0.000 03
g	76	**h**	0.1

3 A triangle has sides of length 5 cm, 8 cm and 10 cm. A second triangle, that is similar to the first, has a perimeter of 92 cm. What are the side lengths of this second triangle?

4 Tony calculates that he will need 5 bags of fertiliser to fertilise his front lawn which is rectangular in shape and has dimensions 5 metres × 8 metres.

How much fertiliser would be needed to fertilise a rectangular lawn that is three times as long and three times as wide as Tony's, assuming the same rate of application?

5 Determine the better buy for the butter deals shown on the right.

250 g
$4.85

100 g
$2.10

6 A map has a scale of 1 : 50 000, i.e. 1 cm on the map represents 50 000 cm (= 500 metres) on the ground.

    a On the map, the straight line distance between two locations is 154 mm. How far apart are these two locations in real life?

    b If two locations are 2 km apart in real life, how far apart are they on the map?

    c On the map the grounds of a shopping centre occupies an area of 1 cm^2. What area does this shopping centre occupy in real life?

7 What is the ratio of the lengths of corresponding sides in two similar triangles given that the ratio of the areas of these two triangles is 49 : 25?

8 Which earns more interest and by how much more:

$2000 invested for 4 years at 8% per annum compounded six monthly,

or

$1800 invested for 5 years at 6% per annum compounded monthly?

9 For the matrices A = $\begin{bmatrix} 2 & 0 \\ 1 & -1 \end{bmatrix}$, B = $\begin{bmatrix} 1 & 3 \end{bmatrix}$ and C = $\begin{bmatrix} -1 \\ 2 \end{bmatrix}$ form all of the products DE that can be formed given that D can be chosen from A, B or C and E can be chosen from A, B or C.

10 A reduced-size scale model of a vehicle is made to a scale of 2 : 35. If the real vehicle is of length 3.85 metres how long is the scale model?

11 Three shops each have special deals on a particular brand of chocolate.

Shop A: Buy two 250 gram bars, normally $4.20 each, and we will give you 80 cents off each bar.

Shop B: Buy ten 200 gram bars at $2.95 each bar and we will give you two extra 200 gram bars free.

Shop C: 15% off the normal cost of a 500 gram bar. (Normal cost $7.80.)

Rank the three deals, best value for money first.

12 A map has a scale of 1 : 500 000.

    a The distance between two locations on the map is 6.2 cm. How far are these two locations apart on the ground?

    b A forestry area occupies 0.8 cm^2 on the map. What area does this forestry area occupy in real life?

13 A solid metal cube of side length 10 cm is melted down and the metal is recast into cubes of side length 1 cm.

Assuming no metal is lost in this process:

    a how many of the smaller cubes will there be?

    b how does the total surface area of all the smaller cubes compare with the surface area of the original cube?

14 Let us suppose that a particular government child education allowance gives, to each family, $250 per month per child in full time education and living at home. However, the total monthly amount received under this scheme reduces by $0.50 for each dollar that the combined parental earned income exceeds $850 per week.

For what combined parental weekly earned income does this allowance cut out (i.e. reduce to zero) for a family with

   a   1 child?   **b**   2 children?   **c**   3 children?

15 How many rectangular blocks with dimensions a cm × b cm × c cm can fit into a rectangular space that is $5a$ cm × $5b$ cm × $5c$ cm?

16 If it takes 16 seconds for a machine to pump up a balloon of radius 30 cm how long does this suggest it would take the machine to pump up a balloon of radius 45 cm?

17 Two spheres, A and B, are such that the ratio of the surface area of sphere A to that of sphere B is 1 : 16. If sphere A has a volume of 50 cm^3 what is the volume of sphere B?

18 How much sodium chloride should be added to each of the following containers, assumed 80% full of water, to give a concentration of 3 grams of sodium chloride per litre of water. (Assume that the given measurements are internal dimensions.)

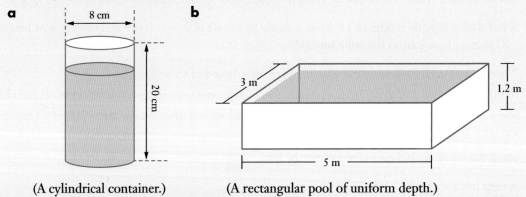

               (A cylindrical container.)   (A rectangular pool of uniform depth.)

19 An 8 metre ladder is placed with its foot on horizontal ground and its top just reaching the top of a vertical wall of height 7 metres. With the ladder remaining in contact with both the wall and the ground, the base of the ladder is pulled a further 1 metre from the wall. How far does this action cause the top of the ladder to move down the wall?

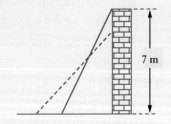

20 Jen works at a garden centre and is given the job of polishing the two large solid stone spheres that proudly adorn the entrance to the centre, one of the spheres being twice the volume of the other.

Jen starts on the smaller one and finds that it takes her 10 minutes to polish it.

How long would this suggest it would take Jen to polish the larger one?

Explain any assumptions you make in calculating your answer.

ANSWERS

Notes:

- For questions that do not stipulate a specific level of rounding the answers given here have been rounded to a level considered appropriate for the question.

- If a question asks for an answer to be given 'to the nearest millimetre' it does not necessarily have to be given 'in millimetres'. In such a situation an answer of 14.768 cm could be written as 14.8 cm or as 148 mm, both answers being to the nearest millimetre.

Exercise 1A   PAGE 7

1 a $v = 23$ **b** $v = 22$ **c** $v = 52$

2 a $C \approx 18.8$ **b** $C \approx 94.2$ **c** $C \approx 17.6$

3 a $V \approx 113$ **b** $V \approx 905$ **c** $V \approx 4189$

4 a $s = 40$ **b** $s = 36$ **c** $s = 133$

5 a $A \approx 705$ **b** $A \approx 990$ **c** $A \approx 1242$

6 a $C = 100$

  b The cost of producing one kg of the metal of 99% purity is approximately $1000.

  c The cost of producing one kg of the metal of 99.9% purity is approximately $10 000.

7 a $C = 9800$

  b The cost of producing one tonne of the metal of 75% purity is approximately $11 000.

  c The cost of producing one tonne of the metal of 95% purity is approximately $23 800.

8 a At a depth of 5 metres in liquid of density 1000 kg/m^3 the pressure is 49 000 N/m^2.

  b At a depth of 10 metres in liquid of density 1030 kg/m^3 the pressure is 100 940 N/m^2.

  c At a depth of 30 metres in liquid of density 1030 kg/m^3 the pressure is 302 820 N/m^2.

9 a i To the nearest half unit the dose for a child aged 5 years would be 4.5 units.

   ii To the nearest half unit the dose for a child aged 10 years would be 8.5 units.

   iii To the nearest half unit the dose for a child aged 15 years would be 12.5 units.

  b The rule makes sense for c up to 18, after which the formula would be giving the 'child' more than the adult dose.

10 a The density of aluminium is 2.7 g/cm^3.

  b The density of lead is 11 350 kg/m^3.

  c The density of diamond is 3.51 g/cm^3.

  d Just below the surface sea water has a density of 1.0282 g/cm^3.

  e At a depth of 1000 m sea water has a density of 1.0328 g/cm^3.

  f At a depth of 10 000 m sea water has a density of 1.071 g/cm^3.

11 a The amount in the account after 5 years is $483.15.

  b The amount in the account after 5 years is $881.17.

  c The amount in the account after 5 years is $16 830.62.

12 a The value of T is 2.15 (correct to 2 decimal places).

  b The value of T is 1.27 (correct to 2 decimal places).

  c The value of T is 0.90 (correct to 2 decimal places).

Spreadsheets   PAGE 10

Compare the output from your spreadsheets with those of others in your class.

Miscellaneous exercise one   PAGE 12

1 a 16 **b** 14 **c** 17 **d** 100

  e 26 **f** −1 **g** 1 **h** 3

  i 1.414 **j** 2.02 **k** 10 **l** 8

2 a 7 **b** 10 **c** 25 **d** 13

  e 17 **f** 125 **g** 19 **h** −19

  i 16 **j** 64 **k** 8 **l** 28

  m 4 **n** 6 **o** 5 **p** 4

  q 6 **r** 2

3 a $208 \times 84 \approx 200 \times 80$. Hence a reasonable estimate would be 16 000.

However, with multiplication a better estimate would be obtained if we rounded one number up and the other down (by a similar proportion).

Using this idea $208 \times 84 \approx 200 \times 90$ giving an estimate of 18 000.

(Check on your calculator which estimate is closest to the accurate answer.)

b $19.6 \times 4.7 \approx 20 \times 5$. Hence a reasonable estimate would be 100.

However, with multiplication a better estimate would be obtained if we rounded one number up and the other down.

Using this idea $19.6 \times 4.7 \approx 20 \times 4.5$ giving an estimate of 90.

c $\frac{208}{9.7} \approx \frac{200}{10}$. Hence a reasonable estimate would be 20.

However, with division a better estimate would be obtained if we rounded both numerator and denominator the same way (by a similar proportion).

Using this idea $\frac{208}{9.7} \approx \frac{210}{10}$ giving an estimate of 21.

d Using the thinking of the previous part a reasonable estimate would be $\frac{5000}{100}$ i.e. 50 and a better estimate would be $\frac{4800}{100}$ i.e. 48.

e $623 \times 80 \text{ cm} \approx 600 \times 80 \text{ cm}$
$= 48\,000 \text{ cm i.e. about 500 metres.}$

4 a Julie is considered to be the *correct* weight.

b Alex is considered to be *under*weight.

c Bill is considered to be *over*weight.

d Betty is considered to be the *correct* weight.

Exercise 2A   PAGE 19

1 a 0.1   **b** 0.3   **c** 0.25   **d** 0.04
   e 0.125   **f** 1.4   **g** 1.1   **h** 1.23
   i 1.04   **j** 1.125   **k** 0.9   **l** 0.92
   m 0.82   **n** 0.4   **o** 0.975

2 a 21 students out of 50 is 42%.

b $18 out of $25 is 72%.

c $2.25 out of $18 is 12.5%.

d 174 sheep out of 1356 sheep is 12.8%, to nearest 0.1%.

e 8.5 cm out of 2.5 metres is 3.4%.

f 35 metres out of 5.832 kilometres is 0.6%, to nearest 0.1%.

3 a $20   **b** $60   **c** $6
   d $22   **e** 40 kg   **f** $12.65
   g 0.6 metres (= 60 cm)
   h 1.35 tonnes (= 1350 kg)

4 a $60   **b** $96
   c 176 kg   **d** 77 metres
   e $42.21   **f** $585
   g 63 litres   **h** $281.25

5 a $40   **b** $13.50
   c 405 kg   **d** 5.88 metres (= 588 cm)
   e 5.2 metres   **f** $2.30
   g $77   **h** 88 tonnes

6 a i $28.66   **ii** $28.65   **iii** $28.70
   b $1443   **c** $1077

7 a The amount is $1345.

b The amount is $400.

c The amount is $260.

d The amount is $158.50.

e The amount is $12 540.

f The price of the item before the rise was $244.

g Before the rise the shares were worth $1296 (nearest dollar).

h The normal price of the item is $47.

i The normal price of the item is $128.

j Before the rise Joe's weekly pay was $835.

k The manufacturer sold 2340 cars the previous month.

8 The price of the commodity has increased by 9.7%, correct to 1 decimal place.

9 The share price has decreased by 6.4%, correct to 1 decimal place.

10 There are 14 girls in the class.

11 There are 32 students in the class.

12 Stamp duty payable is $24 830.

		Number of items	Cost per item	Sub total	GST (10%)	Total
13	e.g.	15	$16.40	$246.00	$24.60	$270.60
	a	23	$17.50	$402.50	$40.25	$442.75
	b	131	$16.40	$2148.40	$214.84	$2363.24
	c	18	$15.90	$286.20	$28.62	$314.82
	d	24	$17.50	$420.00	$42.00	$462.00
	e	6	$19.85	$119.10	$11.91	$131.01
	f	15	$75.30	$1129.50	$112.95	$1242.45
	g	26	$8.00	$208.00	$20.80	$228.80
	h	14	$6.75	$94.50	$9.45	$103.95
	i	124	$3.40	$421.60	$42.16	$463.76
	j	18	$38.75	$697.50	$69.75	$767.25

14 The percentage increase is 8%, to the nearest percent.

15 After the rise the person will be earning $2065.77 per fortnight, to the nearest cent.

16 The total rainfall for the region in 2007 was 226 mm, to the nearest millimetre.

17 a The child dose is 10 milligrams.

b The adult dose is 12.5 millilitres.

18 The cost of the house has increased by 50.9%, correct to one decimal place.

19 To one decimal place a percentage increase of at least 4.8% is required to take the fortnightly pay to at least $1500.

20 In the sale you should expect to pay $106.25 for the drill, $277.95 for the chain saw, $30.60 for the sander and $23.80 for the tool box.

21 a Aimee will pay no income tax.
Brittney will pay $2850 income tax.
Chris will pay $3905 income tax.
Devi will pay $11 924 income tax.
Emily will pay $23 898 income tax.
Frank will pay $69 564 income tax.

b Megan has a taxable income of $105 600.

c Allen has a taxable income of $56 800.

22 The region produced approximately 133 000 kg the year before and approximately 123 000 kg the year before that.

23 The pretax cost of the item is $176.00.

Exercise 2B   PAGE 23

1 a $46.53 (or $46.55 rounded to nearest five cents).

b $48.11 (or $48.10 rounded to nearest five cents).

c $49.75.

We would not want to place huge reliability on the predicted prices because inflation rates can vary and may well not remain steady at the 3.4% quoted. Certainly we would not want to claim the accuracy of the nearest five cent prices quoted above. Also, inflation rates are based on a selection of goods. The price of one particular type of commodity may, for some reason, experience price changes out of line with the general market. However, in the absence of further information the predicted values could be 'as good as we can get' to make such estimates and would perhaps allow us to expect a price rise of about $1.50 per year on the item over the next three years.

2 Compare your answer with those of others in your class and with your teacher.

3 a $8795   **b** $84   **c** $2285

Exercise 2C   PAGE 26

1 In the order displayed, left to right, the discounted prices are $46, $61.90, $66.60 and $79.95 (rounding to the nearest 5 cents when necessary).

2 Before the discount the price of the item was $128.

3 A 10% discount is needed to reduce $75 to $67.50.

4 The discounted price of the item will be $44.62.

5 The price of Order One would be $7707.53.
The price of Order Two would be $435.10.
(No discount given as order not over $500.)
The price of Order Three would be $867.53.

6 a $11.20   **b** $2240   **c** 160

7 The agency charges $13 000 for the sale.

8 The sales person earns $3204 for the month.

9 The salesperson earns $4180 for the month.

10 The real estate agent is paid $10 720.

11 The salesperson receives $2294.40 for the fortnight.

12 The total value of the sales the previous fortnight was $18 240.

13 a The commission charged is $3000.

b The commission charged is $3600.

c The commission charged is $8375.

d The commission charged is $19 620.

	What it cost	What it was sold for	Profit as percentage of cost
14	$100	$124	24%
15	$400	$418	4.5%
16	$100	$118	18%
17	$650	$845	30%
18	$125	$135	8%
19	$12 500	$20 625	65%

	What it cost	What it was sold for	Loss as percentage of cost
20	$100	$84	16%
21	$175	$105	40%
22	$6500	$6110	6%
23	$18.50	$14.80	20%
24	$32.50	$29.25	10%
25	$12 100	$11 132	8%

26 Item A shows the greater percentage profit (28.57% compared to 28.35%).

27 Meta paid $750 for the item.

28 Toni sold the item he bought for $85 for between $102 and $119, the item he bought for $155 for between $186 and $217 and the item he bought for $2150 for between $2580 and $3010.

29 Jack paid $2500 for the item.

Miscellaneous exercise two   PAGE 29

1 **a** 12 **b** 12 **c** 35 **d** 19
  e 17 **f** 24 **g** 54 **h** 175
  i 144 **j** 74 **k** 4 **l** 27

2 **a** 1.2 **b** 0.2 **c** 0.8 **d** 0.02
  e 0.98 **f** 1.02

3 B ($840), E ($702), C ($600), A ($525), D ($448), F ($360).

4 25%.
  Yes the question can be answered without knowing the quantities '$17.50' and '200', as follows:
  100 units of normal cost, with 20% discount, becomes an 80 unit expenditure. Selling an 80 unit expenditure for 100 units gives a profit of 20 units on an expenditure of 80 units, i.e. 25% profit.

5 There are at least 288 year 8 students in the school but no more than 303.

6 His selling price should be $22.

7 No.
  $1.05 \times 1.05 \times 1.05 \times 1.05 \times 1.05 \times 1.05 \times 1.05 \times 1.05 \times 1.05 \times 1.05 = 1.6289$ rounded to 4 decimal places, i.e. after ten years the increase in the cost of living will be approximately 63%, not 50%.

8 **a** 403 **b** 910 **c** 49% (to nearest percent)

9 **a** **i** $114 **ii** $2422.50
  iii $12 065 **iv** $35 940

  b
Dutiable value	Stamp duty payable
$0 to $150 000	1.5% of dutiable value.
$150 001 to $300 000	$2250 + 2.8% of dutiable value over $150 000
$300 001 to $500 000	$6450 + 3.5% of dutiable value over $300 000
$500 001 to $750 000	$13 450 + 4.2% of dutiable value over $500 000
Over $750 000	$23 950 + 5.1% of dutiable value over $750 000

10 Jill paid $324 000 for the house and Jack paid $180 000 for it.

11 $V \approx 15.05$ m^3 (2 decimal places). The container can safely hold 6.02 m^3 (or less).

Exercise 3A   PAGE 36

1 Interest paid is $400.

2 $360 in interest is earned and the final value of the investment is $860.

3 After 15 years the account will be worth $8600.

4 At closure the account will be worth eleven thousand seven hundred and forty five dollars.

5 **a** Three years later the account will be worth $2905.70.

  b The account would be worth $1186 more. (I.e. the extra $1000 invested + extra interest of $186.)

6 He would have received $145.80 more.

7 $4500 interest is earned.

8 $169 863.01, to the nearest cent.

9 The account will be worth $792.19, to the nearest cent.

10 $371.58, to the nearest cent.

11 The account will be worth $53 807 at the end of this time.

12 The special offer rate will give her an extra $45.48, to the nearest cent.

13 $329.65 to the nearest cent.

ISBN 9780170390194

Exercise 3B PAGE 40

1 August $1.29
      September $1.83
      October $6.89

2 March $7.95
      (Lowest balance for March being on 1 to 6 March.)
      April $10.55
      May $27.02

3 July $25.74   August $25.74
      September $25.74   October $25.74
      November $26.78   December $26.78
      January $26.78   February $26.78
      March $11.46   April $3.58
      May $3.58   June $19.34

4 April $6.32 (= $1.4824 + $2.5347 + $0.6039
                  + $1.7010)
      May $9.95 (= $2.2680 + $0.7314 + $3.2089
                  + $0.3328 + $0.8301 + $2.5754)

5 $16.67 (= $2.4612 + $2.4549 + $2.1057 + $5.6589
            + $3.1966 + $0.7925)

Exercise 3C PAGE 42

1 $3 500   **2** $3 675

3 $168.23 to the nearest cent.

4 $480.82 to the nearest cent.

5 $9 960   **6** $23 562.50

7 $385 is owed after three years.

8 Tarni will have to pay $9538 to clear the loan after 3 years.

9 Altogether Frank owes $13 525 at the end of the 3 years.

10 Ali will need to repay $3105 to clear the new loan 2 years after starting it.

Miscellaneous exercise three PAGE 43

1 **a** 35.33   **b** 25.82   **c** 56.97   **d** 27.26

2

	What it cost	What it was sold for	Profit as percentage of cost
a	$200	$250	25%
b	$450	$540	20%
c	$1 650	$2 310	40%

	What it cost	What it was sold for	Loss as percentage of cost
d	$200	$190	5%
e	$8 500	$8 330	2%
f	$156 000	$93 600	40%

3 **a** 25% of $500 is $125.
     b Increasing $500 by 25% gives $625.
     c The amount is $2000.
     d The amount before the increase was $400.

4 **a** B is the odd one out.
     b B is the odd one out.
     c A is the odd one out.
     d C is the odd one out.

5 At the end of the five years the investment has a value of $7076.

6 Neither account pays more interest than the other. They each pay $1400.

7 $A = 204.2$

8 **a** The formula suggests that a vehicle that left a skid mark of 22 metres would have been travelling at approximately 60 km/h.
     b The formula suggests his speed was approximately 90 km/h and thus supports his claim. (However, if the evidence suggested that the car still had significant speed at the end of the skid that would indicate it was going faster than the 90 km/h indicated by the formula.)

9 **a** Rounding can result in the total adding up to more than 100. For example, consider the three numbers 38.47, 26.56 and 34.97.
        Whilst 38.47 + 26.56 + 34.97 = 100, if we round each number to one decimal place and then add up the rounded answers we get a total of 100.1:
        38.5 + 26.6 + 35.0 = 100.1.
     b **i** Approximately 13 000.
          ii Approximately 2000.
     c Assuming that the nickel sector approximately maintains its 12% share of the total, approximately 8000 to 8500 people will be employed in the nickel sector 12 years after the year that the pie chart percentages referred to.

10 He should sell each of the remaining cars for $13 520.

Exercise 4A PAGE 51

(In this chapter some answers may vary slightly dependent upon whether accurate answers are carried forward or 'rounded to nearest cent' answers are carried forward.)

1 The investment will be worth $6077.53 at the end of the four years.

2 After twenty five years the investment is worth $1369.70.

3 After three years $410.76 is owed.

4 **a** $124.86   **b** $126.16   **c** $126.83

5 **a** $2024.64   **b** $2092.60   **c** $2153.84

6 After two years $2254.32 is owed.

7 **a** $2480   **b** $2508.80
   c $2539.47   **d** $2542.40

8 $153 521.73 will need to be repaid on the loan 15 years later.

9

	Simple interest	Compounded annually	Compounded every 6 months	Compounded quarterly
Amount borrowed	$10 000	$10 000	$10 000	$10 000
Amount owed after 1 year	$10 800	$10 800	$10 816	$10 824.32
Amount owed after 2 years	$11 600	$11 664	$11 698.59	$11 716.59
Amount owed after 3 years	$12 400	$12 597.12	$12 653.19	$12 682.42
Amount owed after 4 years	$13 200	$13 604.89	$13 685.69	$13 727.86
Amount owed after 10 years	$18 000	$21 589.25	$21 911.23	$22 080.40
Amount owed after 20 years	$26 000	$46 609.57	$48 010.21	$48 754.39

10

	Simple interest	Compounded annually	Compounded every 6 months	Compounded monthly
Initial balance	$2000	$2 000	$2 000	$2 000
Balance after 1 year	$2240	$2 240	$2 247.20	$2 253.65
Balance after 2 years	$2480	$2 508.80	$2 524.95	$2 539.47
Balance after 3 years	$2720	$2 809.86	$2 837.04	$2 861.54
Balance after 4 years	$2960	$3 147.04	$3 187.70	$3 224.45
Balance after 10 years	$4400	$6 211.70	$6 414.27	$6 600.77
Balance after 20 years	$6800	$19 292.59	$20 571.44	$21 785.11

Exercise 4B PAGE 54

1 **a** When one year old the car will have a value of approximately $28 200.
   b When five years old the car will have a value of approximately $16 900.

2 **a** Two years from now the value of the house will be approximately $385 000.
   b Twenty years from now the value of the house will be approximately $890 000.
   c Fifty years from now the value of the house will be approximately $3 650 000.

3 Assuming a constant annual inflation rate of 4% the chocolate bar will cost approximately $4.80 in 20 years time.

   If instead the inflation rate were 8% the chocolate bar would cost $10.25 in 20 years time.

4 **a** Three years from now the car will have a value of approximately $25 600.
   b Five years from now the car will have a value of approximately $22 000.
   c Ten years from now the car will have a value of approximately $15 000.

5 At the end of the five year period the commodity would cost approximately $103 per kg.

1 a $8 **b** $8 **c** $36
 d $6.96 **e** $870 **f** $352.60

2 a 6 **b** 8 **c** 10 **d** 3.5

3 In ten years time the item will cost approximately $370.

4 Anje should choose scheme B for a value after three years of $10 190.79.

5

Year	Interest for the year	Loan amount
1	$5 100	$65 100
2	$5 533.50	$70 633.50
3	$6 003.85	$76 637.35
4	$6 514.17	$83 151.52
10	$10 627.66	$135 659.01
25	$36 131.33	$461 205.74

6 Tables for 'tax changes announced' not given here. Compare your answers with others in your class.

Exercise 5A PAGE 60

1 $740 **2** $1267.20
3 $1288 **4** $1998.88
5 $809.60 **6** $1743.30
7 $875
8 a $22.50 **b** $67.50
9 $1210 **10** $872

11 Compare your answer with those of others in your class and check with your teacher.

12 a $5750 **b** $1250 **c** $2153.85
 d $40 560 **e** $67 080

13 $8000 per month.
$1680 per week.
$86 000 per annum.
$7000 per month.
$3210 per fortnight.
$41.20 per hour, 38 hour week, no overtime.
$38.75 per hour, 40 hour week, no overtime.
$75 000 per year.

Exercise 5B PAGE 65

1 a $0.0625/g **b** $0.093/g
 c $0.0336/g **d** $0.0168/g
2 a $5.93/100 g **b** $6.47/100 g
 c $1.29/100 g **d** $3.70/100 g

3 a $24/kg **b** $23.20/kg
 c $20/kg **d** $15/kg

4 3 L for $6.75 is $2.25 per litre.
1.8 L for $4.50 is $2.50 per litre.
2.4 L for $5.40 is $2.25 per litre.
On this basis the 3 L for $6.75 and the 2.4 L for $5.40 are equally the 'best buys'.

5 400 g for $4.65 is $1.1625/100 g.
600 g for $6.40 is $1.067/100 g (rounded to 3 dp).
1 kg for $10.50 is $1.05/100 g.
On this basis the 1 kg for $10.50 is the 'best buy', then 600 g for $6.40, then 400 g for $4.65.

6 In order, best value first:
375 g for $5.20 (which is $1.387/100 g, rounded to 3 dp).
500 g for $7.00 (which is $1.40/100 g).
250 g for $3.95 (which is $1.58/100 g.).

Exercise 5C PAGE 68

1 a NZ$929 **b** A$1008
2 a ¥769 280 **b** A$83.19
3 a R2760 **b** A$543.54
4 a £558.62 **b** A$7608
5 a RM7975 **b** A$2507.92
6 A$623.95 **7** A$197.10
8 Pete paid out A$1938.92. Pete gets back A$1900.24
9 A$342 **10** A$4020
11 a A$257.40 **b** A$40

Exercise 5D PAGE 72

1 Total value $62 000 Total dividend $7019
2 Total value $819 307 Total dividend $48 990
3 Total value $103 827.50 Total dividend $5127
4 Total value $112 497 Total dividend $4919.74
to nearest cent.

5 12

6 13.47 rounded to 2 decimal places

7 Premiere Bank (7.51)
Japatali Fund (9.56)
Jupiter Trust (11.50)
Tacomala Group (12.43)
DeepGas Ltd (12.49)
Iron Resources (13.82)
Linear Corp (22.18)

8 a 10 **b** 20 **c** 12.5%

9 The share price can depend on a number of things, one of which is likely to be the value of the assets and cash held by the company. Just prior to paying out some of the profits as dividends these profits are owned by the company and hence contribute to the share value. Once the dividends are paid out (i.e. the company goes 'ex-dividend') the company no longer owns this money and so the company value, and hence the share price, will fall to reflect this fall in company value.

Exercise 5E   PAGE 78

1 $3600.40   **2** $4300

3 Nil   **4** $115.38 (nearest cent)

5 $5100

6 $86 to the nearest dollar.

7 $566 to the nearest dollar.

8 $403 to the nearest dollar.

9 $720 to the nearest dollar.

10 $567 to the nearest dollar.

11 $236 to nearest dollar.

12 $570 to the nearest dollar.

Miscellaneous exercise five   PAGE 81

1 a 80% of $55 is the greater.

b 93% of $95 is the greater.

c The two are equal. Neither one is greater than the other one.

d The two are equal. Neither one is greater than the other one.

2 a $5955.08   **b** $5970.26

c $5978.09   **d** $5983.40

3 Based on cost per square metre the 'best value for money first' rank order is:

1st   35 m by 42 m
2nd   21 m by 35 m
3rd   18 m by 52 m
4th   17 m by 42 m

When comparing the values of blocks of land it is very unlikely that 'all other things will be equal' with regards to value for money. If the blocks are in different land developments then closeness to the city, closeness of public transport, views, reputation of local schools etc will all make these 'other things' far from equal. Even if the blocks are on the same land development the views, the closeness of amenities such as shops and parkland, whether the block is a corner block etc will again be things that will be far from equal from one block to another.

4 a $473.60   **b** $368.60   **c** $915

5

	Number of students	Percentage of those doing the unit getting				
		As	Bs	Cs	Ds	Fs
Unit I	25	16	28	40	12	4
Unit II	13	15	15	46	23	0
Unit III	11	27	36	18	9	9
Unit IV	14	14	43	29	14	0
Unit V	15	7	40	27	20	7
Unit VI	22	23	36	23	18	0

Exercise 6A   PAGE 89

1 $A_{4 \times 2}, B_{2 \times 4}, C_{4 \times 1}, D_{4 \times 3}, E_{2 \times 2}, F_{1 \times 3}, G_{3 \times 2}, H_{4 \times 4}$

2 a 4   **b** −4   **c** 7

d 7   **e** 3   **f** 0

3 a Cannot be determined

b $\begin{bmatrix} 3 & -1 \\ 1 & -9 \end{bmatrix}$ **c** $\begin{bmatrix} 1 & -5 \\ 1 & -1 \end{bmatrix}$

d $\begin{bmatrix} 6 \\ 2 \\ -4 \end{bmatrix}$ **e** $\begin{bmatrix} 9 & -3 \\ 6 & 12 \\ 0 & 9 \end{bmatrix}$

f Cannot be determined

g $\begin{bmatrix} 2 & 4 \\ 0 & -8 \end{bmatrix}$ **h** $\begin{bmatrix} 0 & 7 \\ -1 & -3 \end{bmatrix}$

4 a $\begin{bmatrix} 5 & 3 & -1 \\ 1 & 3 & 3 \end{bmatrix}$ **b** $\begin{bmatrix} -1 & -1 & 1 \\ -1 & -5 & -3 \end{bmatrix}$

c $\begin{bmatrix} 3 & 6 & 3 \\ 6 & 3 & 6 \end{bmatrix}$ **d** $\begin{bmatrix} 5 & 4 & -3 \\ 3 & 14 & 9 \end{bmatrix}$

5 a Cannot be determined **b** $\begin{bmatrix} 6 & 12 \\ 3 & 9 \end{bmatrix}$

c $\begin{bmatrix} 8 & 3 & 11 \end{bmatrix}$ **d** Cannot be determined

6 a Cannot be determined **b** $\begin{bmatrix} 6 & 4 & 3 & 0 \\ 2 & 2 & 6 & 6 \\ 1 & 5 & 3 & 4 \end{bmatrix}$

c $\begin{bmatrix} 6 & 2 & 8 \\ 4 & 2 & -6 \\ 0 & 2 & 4 \\ 2 & 0 & 0 \end{bmatrix}$ **d** $\begin{bmatrix} 0 & 14 & -3 & 6 \\ -2 & 4 & 6 & 12 \\ -1 & -5 & 3 & 20 \end{bmatrix}$

7 a No   **b** No   **c** Yes   **d** Yes
   e Yes   **f** No   **g** Yes   **h** No

8 Yes   **9** Yes

10 $\begin{bmatrix} 1 & 2 & -3 \\ 1 & 0 & -2 \end{bmatrix}$

11 a

	P	A	B
Alan	40	20	4
Bob	37	15	14
Dave	47	19	9
Mark	39	21	3
Roger	39	19	16

b

	P	A	B
Alan	10	5	1
Bob	9.25	3.75	3.5
Dave	11.75	4.75	2.25
Mark	9.75	5.25	0.75
Roger	9.75	4.75	4

12

	B	F	FL	G	GG
Centre I	6160	1925	2552	1947	4675
Centre II	3124	1397	1507	1122	2992
Centre III	5555	1617	3102	1408	2970
Centre IV	2409	1034	1672	924	1958

13 $\begin{bmatrix} 3 & 4 & 5 \\ 5 & 6 & 7 \\ 7 & 8 & 9 \end{bmatrix}$ **14** $\begin{bmatrix} 1 & 1 & 1 & 1 \\ 2 & 4 & 8 & 16 \\ 3 & 9 & 27 & 81 \end{bmatrix}$

Exercise 6B   PAGE 95

1 $\begin{bmatrix} 4 & 9 \end{bmatrix}$

2 Cannot be determined. Number of columns in 1st matrix ≠ number of rows in 2nd matrix

3 $\begin{bmatrix} 2 & 10 \\ 1 & 4 \end{bmatrix}$ **4** $\begin{bmatrix} 7 \end{bmatrix}$

5 $\begin{bmatrix} 3 & 1 \\ 12 & 4 \end{bmatrix}$ **6** $\begin{bmatrix} 13 & -4 \\ -14 & 7 \end{bmatrix}$

7 $\begin{bmatrix} 2 & 3 \\ 1 & -1 \end{bmatrix}$ **8** $\begin{bmatrix} 1 & 4 \\ -1 & 3 \end{bmatrix}$

9 $\begin{bmatrix} 0 & 0 \\ 0 & 0 \end{bmatrix}$ **10** $\begin{bmatrix} 1 & 0 \\ 0 & 1 \end{bmatrix}$

11 $\begin{bmatrix} 1 & 0 \\ 0 & 1 \end{bmatrix}$ **12** $\begin{bmatrix} 1 & 0 \\ 0 & 1 \end{bmatrix}$

13 $\begin{bmatrix} 8 \end{bmatrix}$ **14** $\begin{bmatrix} 3 & 2 & 3 \\ 4 & 3 & 1 \end{bmatrix}$

15 $\begin{bmatrix} 1 & 0 & 5 \\ 10 & 2 & -2 \\ 6 & 1 & 4 \end{bmatrix}$ **16** $\begin{bmatrix} 10 & 3 \\ 9 & 10 \end{bmatrix}$

17 $\begin{bmatrix} 14 \\ 32 \end{bmatrix}$ **18** $\begin{bmatrix} 2 & 4 & 1 \\ 5 & 7 & 18 \\ 12 & 8 & 22 \end{bmatrix}$

19 a $\begin{bmatrix} 0 & 2 & 1 \\ 0 & 1 & 5 \\ 2 & 0 & 1 \end{bmatrix}$ **b** $\begin{bmatrix} 2 & 2 & 3 \\ 4 & 0 & -1 \\ -2 & 1 & 0 \end{bmatrix}$

   c $\begin{bmatrix} 1 & -1 & -2 \\ 2 & 1 & -1 \\ 2 & 1 & 2 \end{bmatrix}$ **d** $\begin{bmatrix} 2 & -1 & 2 \\ 2 & 3 & 4 \\ -2 & -2 & 1 \end{bmatrix}$

20 No. Justify by showing example for which AB ≠ BA

24 a Cannot be formed   **b** Cannot be formed
   c 3×3 **d** 2×2
   e Cannot be formed   **f** 1×2
   g 3×2 **h** 1×3

25 a Yes   **b** Yes   **c** Yes   **d** No
   e No   **f** No   **g** No   **h** Yes

26 Matrix A must be a square matrix.

27 AA, AC, BA, CB

28 a $\begin{bmatrix} -1 & -2 \\ 4 & 0 \end{bmatrix}$ **b** $\begin{bmatrix} 2 & -2 \\ 7 & -3 \end{bmatrix}$

29 a 1st B, 2nd E, 3rd C, 4th D, 5th A
   b 1st =B & C, 3rd E, 4th D, 5th A

30 Initially: $\begin{matrix} \text{Client 1} \\ \text{Client 2} \\ \text{Client 3} \end{matrix} \begin{bmatrix} \$15\,000 \\ \$15\,000 \\ \$15\,000 \end{bmatrix}$

Two years later: $\begin{matrix} \text{Client 1} \\ \text{Client 2} \\ \text{Client 3} \end{matrix} \begin{bmatrix} \$17\,700 \\ \$19\,300 \\ \$18\,800 \end{bmatrix}$

31 $\begin{matrix} \text{Drinks (mLs)} & \text{Burgers} \\ \begin{bmatrix} 18\,125 & 55 \end{bmatrix} \end{matrix}$

32 a QP
   b $\begin{matrix} \text{Hotel A} & \text{Hotel B} & \text{Hotel C} \\ \begin{bmatrix} \$4610 & \$3680 & \$2665 \end{bmatrix} \end{matrix}$

     Displays total nightly tariff for each hotel when full.

c Row 1 column 1 of PR would be

Single rooms in A × Single room tariff +

Single rooms in B × Double room tariff +

Single rooms in C × Suite tariff

Thus PR not giving useful information.

33 a $\begin{bmatrix} 3 & 1 & 2 \end{bmatrix}$

b
Poles	Decking	Framing	Sheeting
25	205	145	320

Matrix shows number of metres of each size of timber required to complete order.

c $\begin{bmatrix} \$4 \\ \$2 \\ \$3 \\ \$1.50 \end{bmatrix}$

Product will have dimensions 3 × 1.

Matrix will display the total cost of timber for each type of cubby.

34 a
$$\begin{array}{ccc} & A & B & C \\ E = [& 800 & 50 & 1000 &] \end{array}$$

b
Model I	Model II	Model III	Model IV
4600	4900	6300	5600

Matrix displays the total cost of commodities, in dollars, for each model type.

35 a RP

b $\begin{bmatrix} 6700 & 7200 & 2300 \end{bmatrix}$

c Matrix shows the number of minutes required for cutting (6700 minutes), assembling (7200 minutes) and packing (2300 minutes) to complete the order.

Exercise 6C PAGE 104

1 a $B = \begin{bmatrix} -2 & 0 \\ 4 & 3 \end{bmatrix}$ **b** $C = \begin{bmatrix} -1 & 0 \\ 4 & 4 \end{bmatrix}$

2 a $E = \begin{bmatrix} 0 & 0 \\ 0 & 0 \end{bmatrix}$ **b** $F = \begin{bmatrix} 5 & -1 \\ 2 & 0 \end{bmatrix}$

c $G = \begin{bmatrix} 6 & -1 \\ 2 & 1 \end{bmatrix}$ **d** $H = \begin{bmatrix} 5 & -1 \\ 2 & 0 \end{bmatrix}$

e $J = \begin{bmatrix} 5 & -1 \\ 2 & 0 \end{bmatrix}$

3 a True **b** True
c Not necessarily true. **d** True
e True **f** True
g True **h** True
i Not necessarily true. **j** Not necessarily true.

4 a
	A	B	C
A	0	1	1
B	1	0	1
C	1	1	0

b
	A	B	C
A	0	1	0
B	1	1	1
C	0	0	0

c
	A	B	C
A	2	1	0
B	1	0	2
C	0	1	0

5 a
	A	B	C	D
A	0	1	1	1
B	1	0	1	0
C	1	1	0	0
D	1	0	0	0

b
	A	B	C	D
A	0	1	0	0
B	1	0	1	2
C	0	1	2	0
D	1	2	0	0

c
	A	B	C	D
A	0	1	2	0
B	1	0	2	0
C	2	2	0	1
D	1	1	1	0

6 a
	A	B	C	D	E
A	0	1	0	0	1
B	1	0	1	1	1
C	0	1	0	1	0
D	0	1	1	0	1
E	1	1	0	1	0

b
	A	B	C	D	E
A	0	1	0	0	1
B	1	0	1	0	0
C	0	1	0	1	0
D	0	0	1	0	1
E	1	0	0	1	0

c

	A	B	C	D	E
A	0	1	0	1	1
B	1	0	1	0	0
C	0	1	2	0	0
D	1	1	0	0	1
E	1	0	0	0	0

7 The two stage route matrices are as follows.

a

	A	B	C
A	1	0	1
B	0	2	0
C	1	0	1

b

	A	B	C
A	3	2	1
B	1	2	2
C	1	1	3

c

	A	B	C
A	3	3	2
B	5	6	3
C	2	4	6

d

	A	B	C	D
A	2	0	2	0
B	0	2	0	2
C	2	0	2	0
D	0	2	0	2

e

	A	B	C	D
A	2	2	2	1
B	1	3	2	2
C	2	1	2	0
D	0	2	1	2

f

	A	B	C	D
A	5	0	0	2
B	0	1	2	0
C	0	2	5	0
D	2	0	0	1

9

	Sue Min	Tanya	Julie	Peta	Donelle	Mandy
Sue Min	0	1	1	0	1	0
Tanya	1	0	0	0	1	0
Julie	0	0	0	0	1	0
Peta	0	0	1	0	1	0
Donelle	1	1	1	1	0	1
Mandy	1	0	0	0	1	0

10 a Because if F has been to the movies with G then it follows automatically that G has been with F. It is impossible for F to go with G without G also going with F. Hence no 'one way paths'.

b The matrix will be symmetrical with $A_{ij} = A_{ji}$.

c

	Ann	Bill	Chris	Dave	Enya
Ann	0	0	1	0	1
Bill	0	0	0	1	1
Chris	1	0	0	1	1
Dave	0	1	1	0	1
Enya	1	1	1	1	0

d

	Ann	Bill	Chris	Dave	Enya
Ann	0	1	0	0	1
Bill	1	0	1	0	1
Chris	0	1	0	0	0
Dave	0	0	0	0	1
Enya	1	1	0	1	0

11 Being able to display social interactions using numbers in a matrix allows the situation to be subjected to mathematical manipulation and analysis.

12 a

	Jack	John	Sue	Bill	Ken	Tony	Mary
Jack	0	1	1	1	0	1	0
John	1	0	0	1	0	0	0
Sue	0	0	0	1	0	1	0
Bill	1	1	1	0	1	1	1
Ken	0	0	0	1	0	0	0
Tony	1	0	1	0	0	0	1
Mary	0	0	0	1	0	1	0

b

	Jack	John	Sue	Bill	Ken	Tony	Mary
Jack	0	1	2	2	1	2	2
John	1	0	2	1	1	2	1
Sue	2	1	0	0	1	1	2
Bill	2	1	2	0	0	3	1
Ken	1	1	1	0	0	1	1
Tony	0	1	1	3	0	0	0
Mary	2	1	2	0	1	1	0

Part **a** answer squared is same as part **b** answer except for the leading diagonal.

c Only Tony who does not have Ken's, nor can he get Ken's in a two stage process.

In both one stage and two stage matrices there is a zero in the 'Tony row, Ken column' location (and this is not on the leading diagonal).

13 a One stage:

	Mai	Tiny	Tonto	Pronto	Slow
Mai	0	1	0	0	0
Tiny	1	0	0	1	1
Tonto	0	0	0	1	0
Pronto	0	1	1	0	0
Slow	1	0	0	1	0

Two stage:

	Mai	Tiny	Tonto	Pronto	Slow
Mai	0	0	0	1	1
Tiny	1	0	1	1	0
Tonto	0	1	0	0	0
Pronto	1	0	0	0	1
Slow	0	1	1	0	0

b The two stage matrix is not equal to the square of the one stage matrix because the leading diagonal is different and the 'Slow row Tiny column' is different.

The reasons for these differences are as follows:

The 'Slow row Tiny column' entry shows a 2 in the squared matrix because there are two ways that Slow can contact Tiny in two stages; Slow → Mai → Tiny and Slow → Pronto → Tiny. However on the two stage matrix this is shown as a 1 because the question instructed us to use one of only two entries, a 0 to show no contact and a 1 to indicate contact.

The first entry on the leading diagonal of the squared matrix is a 1 because it has allowed the two stage connection Mai → Tiny → Mai whereas our two stage matrix shows 0 as instructed.

The second entry of the leading diagonal of the squared matrix is 2 because it allows the connections Tiny→ Pronto → Tiny and Tiny → Mai → Tiny whereas our two stage matrix shows 0 as instructed. Etc.

Miscellanous exercise six PAGE 109

1 a Cannot be performed.
b Can be performed. 2 rows and 3 columns.
c Cannot be performed.
d Can be performed. 3 rows and 5 columns.
e Cannot be performed.
f Can be performed. 3 rows and 4 columns.
g Cannot be performed.
h Can be performed. 3 rows and 3 columns.
i Cannot be performed.
j Cannot be performed.
k Can be performed. 1 row and 1 column.
l Can be performed. 5 rows and 3 columns.

2 a $\begin{bmatrix} 4 & 0 \\ 0 & 4 \end{bmatrix}$ **b** $\begin{bmatrix} 10 & -4 \\ 6 & -2 \end{bmatrix}$

c $\begin{bmatrix} 6 & -4 \\ 6 & -6 \end{bmatrix}$ **d** $\begin{bmatrix} 1 & 0 \\ 0 & 1 \end{bmatrix}$

e $\begin{bmatrix} 12 & 0 \\ 0 & 12 \end{bmatrix}$ **f** $\begin{bmatrix} 24 & -16 \\ 24 & -24 \end{bmatrix}$

3 a $1003.20 **b** $1103.52
c $1320 **d** $1219.68

4 a $1476 **b** $1600.64

	$25 000 borrowed at 9% per annum			
	Simple interest	Compounded annually	Compounded 6 monthly	Compounded quarterly
5 Initial amount borrowed	$25 000	$25 000	$25 000	$25 000
Amount owed after 1 year	$27 250	$27 250	$27 300.63	$27 327.08
Amount owed after 2 years	$29 500	$29 702.50	$29 812.97	$29 870.78
Amount owed after 3 years	$31 750	$32 375.73	$32 556.50	$32 651.25
Amount owed after 4 years	$34 000	$35 289.54	$35 552.52	$35 690.54
Amount owed after 10 years	$47 500	$59 184.09	$60 292.85	$60 879.72
Amount owed after 20 years	$70 000	$140 110.27	$145 409.11	$148 253.63

Exercise 7A PAGE 116

1 **a** BC **b** EF **c** GH
  d JL **e** NO **f** QR

2 Statements III and VI are true.

3 Statements III, V and VI are true.

4 Statements I, V and VI are true.

5 $x = 10$ 6 $x = 26$

7 $x = 7$ 8 $x = 15.1$ (1 dp)

9 $x = 16.4$ (1 dp) 10 $x = 3.9$ (1 dp)

11 $x = 14.2$ (1 dp) 12 $x = 10.2$ (1 dp)

13 $x = 24$ 14 $x = 13.0$ (1 dp)

15 $x = 17.1$ (1 dp) 16 $x = 5.8$ (1 dp)

17 $x = 30.2$ (1 dp) 18 $x = 7.4$ (1 dp)

19 $x = 3.3$ (1 dp) 20 $x = 15$

21 $x = 6.9$ (1 dp) 22 $x = 2.9$ (1 dp)

23 $x = 11.6$ (1 dp) 24 $x = 3.1$ (1 dp)

25 AC is of length 59 mm, to the nearest millimetre.

26 RQ is of length 16.1 cm, to the nearest millimetre.

27 XZ is either of length 10.1 cm or of length 14.3 cm, to the nearest millimetre.

28 The longer diagonal exceeds the shorter diagonal by 94 mm, to the nearest millimetre.

Exercise 7B PAGE 120

1 A piece of timber of length 289 cm is needed (to the nearest centimetre).

2 The boat is then 5.8 km from the harbour.

3 The boat travelled 2.8 km due East.

4 The television screen is of size 70 cm.

5 The foot of the ladder is 3.4 metres from the base of the wall (to the nearest 10 cm).

6 The height of the wall is 4.9 metres, to the nearest 10 centimetres.

7 The length of each wire will be 19 metres, rounded to the next metre.

8 The new road will reduce the journey by 11.6 km (correct to one decimal place).

9 The height of the container is 87 cm, to the nearest centimetre.

10 The shortcut makes the journey 31 metres shorter, to the nearest metre.

11 Each rectangular side of the tent has an area of 6 m².

12 Each side of the square is of length 85 mm, to the nearest millimetre.

13 The length of steel needed for the 100 frames is, rounded up to the next whole metre, 435 m.

14 The perimeter of the trapezium is 134.2 metres, correct to one decimal place.

15 The length of BC is 4.7 metres, correct to one decimal place.

16 Layout B uses the smaller total length of piping, by 69 cm (to the nearest cm).

17 The area of the triangle is, to the nearest ten square metres, 8380 m².

18 For the 3 m × 1.4 m frame:
Design A requires the greater length of steel by 1.12 metres.

For the 1.8 m × 1.4 m frame:
Design B requires the greater length of steel by 0.17 metres.

Miscellaneous exercise seven PAGE 123

1 \$977.50

2 **a** $\begin{bmatrix} 2 & -4 \\ 3 & -6 \end{bmatrix}$ **b** $\begin{bmatrix} -4 \end{bmatrix}$

3 **a** \$532.35 **b** \$315.00 **c** \$286.65
  d \$585 **e** \$122.85

4 BAC $\begin{bmatrix} 5 & 5 & 2 & 4 \end{bmatrix}$

5 The longest pole that could fit into the given container is, to the nearest centimetre, 5.74 m.

6 The water slide is 5.25 m long, to the nearest centimetre.

7 First check there are no odd numbers on the leading diagonal.

Then check where the symmetry 'breaks down'.
i.e. where $a_{ij} \neq a_{ji}$.

From these non-symmetrical entries the one way roads can be determined.

8 **a** 276 **b** 66% **c** 27%

9 **a** \$2350 **b** \$10 317 235.01
  c \$14 457 828.11 **d** \$41 112 576.17

Exercise 8A PAGE 128

1 **a** 38 m **b** 80 m²

2 **a** 671 mm **b** 236 cm²

3 **a** 30 m **b** 32 m²

4 **a** 33.7 cm **b** 79 cm²

5 **a** 18.67 m **b** 15.61 m²

6 **a** 34 m **b** 44 m²

7 **a** 95.7 cm **b** 537 cm²

8 **a** 20.57 m **b** 23.43 m²

9 80.5 cm^2 **10** 63 m^2

11 42 cm^2 **12** 75.1 m^2

13 $85\,978 \text{ cm}^2$ **14** 19.43 m^2

15 $20\,094 \text{ cm}^2$ **16** $78\,610 \text{ mm}^2$

Exercise 8B   PAGE 130

1 a $952   **b** $2040, $3145   **c** $486

2 a 6×2.4 m with an excess of 1.6 m.

5×3 m with an excess of 2.2 m.

Minimum excess: 3×2.4 m and 2×3 m with an excess of 0.4 m.

b 8×2.4 m with an excess of 1 m.

7×3 m with an excess of 2.8 m.

Minimum excess: 4×2.4 m and 3×3 m with an excess of 0.4 m.

c 15×2.4 m with an excess of 0.6 m.

12×3 m with an excess of 0.6 m.

Minimum excess:

     1×2.4 m and 11×3 m with no excess.

Or 6×2.4 m and 7×3 m with no excess.

Or 11×2.4 m and 3×3 m with no excess.

3 a $5901.50   **b** $865.80

c $3052.50   **d** $2131.20

e $1731.60   **f** $2238.50

g $5291.00   **h** $666.00

4 a $36036   **b** $6586

c $2554   **d** $5356

5 a

Qty	Item	Unit price ($)	Total ($)
90	Fence post	10.00	900.00
90	Post pack (cement, brackets and nails for 1 post)	18.00	1620.00
1	Gate pack (gate, gate posts, latch, cement and all fastenings)	240.00	240.00
132	3.6 metre rail	8.00	1056.00
		Sub total	3816.00
		GST (10%)	381.60
		Grand total	4197.60

b

Qty	Item	Unit price ($)	Total ($)
143	Fence post	10.00	1430.00
143	Post pack (cement, brackets and nails for 1 post)	18.00	2574.00
1	Gate pack (gate, gate posts, latch, cement and all fastenings)	240.00	240.00
213	3.6 metre rail	8.00	1704.00
		Sub total	5948.00
		GST (10%)	594.80
		Grand total	6542.80

6 $43\,200$ kg

7 a $12\,870\,000$ construction costs, $263\,250$ annual maintenance.

b $10\,296\,000$ construction costs, $210\,600$ annual maintenance.

c $13\,728\,000$ construction costs, $280\,800$ annual maintenance.

d $29\,744\,000$ construction costs, $608\,400$ annual maintenance.

8 a Seed: 2160 kg. Harvest: 48\,600 kg

b Seed: 12\,000 kg. Harvest: 270\,000 kg

c Seed: 1790 kg (nearest 10 kg). Harvest: 40\,250 kg (nearest 10 kg). Approx 40 tonnes

d Seed: 110 kg (nearest 10 kg). Harvest: 2450 kg (nearest 10 kg). Approx 2.5 tonnes

9 a $1194   **b** $281   **c** $244   **d** $232

10 a 4000 follicular units   **b** 5660 follicular units

c 8250 follicular units   **d** 6810 follicular units

11 a 'per linear' metre makes it clear that the cost of the bull nose edging is based on length of edge.

b i

Cost of benchtop	$810.24
Cost of edging	$375.00
Total cost	$1185.24
Quote	$1185

ii

Cost of benchtop	$3689.69*
Cost of edging	$834.16
Total cost	$4523.85
Quote	$4524

*Includes 30% loading for semicircle.

12 [Total area of the eight walls is 89.8 m², to nearest 0.1 m².]

Need 17 litres. Could get 10 L + 4 L + 2 L + 1 L for $359.30 but if not concerned about having more paint left over get 10 L + 2 × 4 L for just $335.80. As well as being cheaper this latter option also has the advantage that if the painting requires a little more paint than the estimate suggests you have paint available.

	Maynard Waters	**Plymptain by Sea**	**Woodstock Valley**
13 a	$226 000	$242 000	$291 000
b	$210 000	$226 000	$271 000
c	$386 000	$414 000	$498 000
d	$300 000	$322 000	$386 000
e	$291 000	$312 000	$375 000

14 [Floor area calculation gives 125.917 m².]

		Finish	
Construction type	**Basic**	**Basic Plus**	**Deluxe**
3 bedroom brick veneer standard design	$136 000	$160 000	$204 000
3 bedroom full brick unique design	$179 000	$210 000	$268 000
4 bedroom brick veneer standard design	$152 000	$179 000	$228 000
4 bedroom full brick unique design	$200 000	$235 000	$300 000

15 a $6565 (nearest dollar)

  b $5442

  c $17 385 (nearest dollar)

Exercise 8C   PAGE 145

1 256 cm²   **2** 32 cm

3 26 cm   **4** 3 m

5 3.09 m, to the nearest centimetre.

6 121 mm, to the nearest millimetre.

7 9.55 m, to the nearest centimetre.

8 319 mm

9 $a = 3, b = 2, c = 6, d = 7, e = 9$.
20 cm², 24 cm², 27 cm², 28 cm², 30 cm², 36 cm².

10 $a = 12, b = 6, c = 8, d = 3.2, e = 6$.

11 $r_3 = 2r_1$, $r_2 = \sqrt{2}r_1$

Miscellaneous exercise eight   PAGE 148

1 a $17 500   **b** $15 700   **c** $5500

2 a $\begin{bmatrix} 1 \\ -3 \end{bmatrix}$ **b** $\begin{bmatrix} -1 & 2 \end{bmatrix}$

  c $\begin{bmatrix} 3 & -2 \\ 15 & -10 \end{bmatrix}$ **d** $[-7]$

3 a £1314   **b** A$887.10

4 The item would have cost him $99.20 if the discount had been 20%.

5 At 87 cents per 50 g the 250 g of sliced ham for $4.35 is the better buy as the 450 g costs more than 88 cents per 50 g.

6 The third side could be 10 cm in length or it could be $\sqrt{28}$ cm in length.
($\sqrt{28}$ cm = $2\sqrt{7}$ cm or 5.29 cm rounded to two decimal places.)

7 To the nearest whole numbers
  a 4823   **b** 6121   **c** 4301

8 The base of the ladder should be placed 2.6 metres from the base of the wall. (To the nearest 0.1 m.)

9 a The height of the building is 19.6 m.

  b The balloon was 180 m high when the sandwich was dropped.

  c The teeth will take approximately 3.5 seconds to reach the ground.

10 21.5%

11 a XY   **b** $\begin{bmatrix} 420 \\ 410 \\ 430 \end{bmatrix}$

  c $\begin{bmatrix} \text{Commodity cost (\$) to produce one model A} \\ \text{Commodity cost (\$) to produce one model B} \\ \text{Commodity cost (\$) to produce one model C} \end{bmatrix}$

12 a 2 bags   **b** 77 bags   **c** 32 bags

13

PATSY LING

Normal hourly rate	*$19.20*	/h	Normal	$19.20	/h
Week	23		Time and a half	$28.80	/h
			Double time	$38.40	/h

Hours worked		Payment due	
Normal	35		$672.00
Time and a half	4		$115.20
Double time	0		$0.00
		Total	**$787.20**

Troy Marcesi

Normal hourly rate	*$21.40*	/h	Normal	$21.40	/h
Week	23		Time and a half	$32.10	/h
			Double time	$42.80	/h

Hours worked		Payment due	
Normal	35		$749.00
Time and a half	6		$192.60
Double time	4		$171.20
		Total	**$1,112.80**

Exercise 9A PAGE 156

1	1350 cm^2	**2**	158 m^2
3	6700 mm^2	**4**	1008 cm^2
5	2700 cm^2	**6**	840 m^2
7	308 cm^2	**8**	1568 cm^2
9	384 cm^2	**10**	564 cm^2

11 8954 cm^2 (nearest cm^2)
12 70 700 cm^2 (nearest 100 cm^2)
13 1850 cm^2 (nearest 50 cm^2)
14 7500 cm^2 (nearest 100 cm^2)
15 1680 cm^2
16 1253 cm^2 (nearest cm^2)

Exercise 9B PAGE 160

1	216 cm^3	**2**	270 cm^3
3	108 500 mm^3	**4**	16 450 cm^3
5	168 m^3	**6**	160 m^3

7	1780 cm^3	**8**	195 cm^3
9	22 300 cm^3	**10**	198 cm^3
11	16 kilolitres	**12**	240 litres

13 4021 millilitres (to the nearest millilitre)
14 2094 litres (to the nearest litre)
15 8000 cubes with each edge of length 1 cm could be made from the larger cube.

1909 spheres of radius 1 cm could be made from the large cube.
16 The volume of material required is 198 600 mm^3 (rounded up to the next 100 mm^3).

Exercise 9C PAGE 163

1	**a**	2.8 m^3	**b**	15.8 m^3
	c	50.4 m^3	**d**	21.9 m^3
2	**a**	2.47 m^3, 2 m^3	**b**	4.524 m^3, 4 m^3
	c	6.72 m^3, 6 m^3	**d**	5.576 m^3, 5 m^3

ISBN 9780170390194

3 a 45.9 m³, i.e. 45.9 kL, 15 hours and 18 minutes to fill to 10 cm from top.

b 14.4 m³, i.e. 14.4 kL, 4 hours and 48 minutes to fill to 10 cm from top.

c 11.16 m³ (rounded to 2 decimal places), i.e. 11.16 kL, 3 hours and 43 minutes to fill to 10 cm from top.

d 23.4 m³, i.e. 23.4 kL, 7 hours and 48 minutes to fill to 10 cm from top.

4 a i 38.6 m³   **ii** 48.2 m³

b i 14.8 m³   **ii** 19.2 m³

c i 13.2 m³   **ii** 17.2 m³

d i 22.7 m³   **ii** 30.7 m³

5 a Capacity of tank is 4500 litres. Needs approximately 2.25 m³ of concrete.

b Capacity of tank is 40 039 litres (i.e. approximately 40 kL). Needs approximately 9.43 m³ of concrete.

6 a 1854 m²   **b** 53 m²

c 58 653 m²   **d** 81 m²

7 a 218.68 litres   **b** 91.5%

d 73.68 metres

8 a i Approximately 2 570 000 m³

ii Approximately 5 800 000 000 kg, i.e. approximately 5.8 million tonnes.

b i Approximately 2 220 000 m³

ii Approximately 4 990 000 000 kg, i.e. approximately 5.0 million tonnes.

c i Approximately 257 000 m³

ii Approximately 577 000 000 kg, i.e. approximately 0.58 million tonnes.

9 a Earth: Approximately 1.1×10^{12} km³.

b Basketball: Approximately 7238 cm³.

c Wrecking ball: Approximately 382 000 cm³ (i.e. approximately 0.382 m³).

d Moon: Approximately 2.21×10^{10} km³.

e Snooker ball: Approximately 75 800 mm³ (i.e. approximately 75.8 cm³).

f Eyeball: Approximately 5.58 cm³.

10 a i Real weight is approximately 14.5 kg. Cubic weight is 12 kg.

Hence charge by real weight.

Charge is $7.50 + 14 × $4.40 = $69.10

ii Real weight is approximately 6.3 kg. Cubic weight is 12 kg.

Hence charge by cubic weight.

Charge is $7.50 + 11 × $4.40 = $55.90

b i Real weight is 4.423 kg. Cubic weight is 6 kg.

Hence charge by cubic weight.

Charge is $8.40 + 5 × $5.20 = $34.40

ii Real weight is 8.452 kg. Cubic weight is 7.5 kg.

Hence charge by real weight.

Charge is $8.40 + 8 × $5.20 = $50

11 a i 87.96 m²   **ii** 33 litres

b i 45.05 m²   **ii** 17 litres

c i 80.92 m²   **ii** 31 litres

d i 31.48 m²   **ii** 12 litres

12 a 810 cm³   **b** 12 945 cm³

c 2295 cm³   **d** 28 845 cm³

Exercise 9D   PAGE 176

1 a 5 cm   **b** 150 cm²

2 a 4 cm   **b** 64 cm³

3 The sphere has a radius of 26 mm, to the nearest millimetre.

4 The radius of the hemisphere is 84 mm, to the nearest millimetre.

5 The radius of the sphere is 247 mm, to the nearest millimetre.

6 The cube has a surface area of 1350 cm².

7 The sphere has a surface area of 1088 cm² (nearest cm²).

8 $v = 6$, $w = 8$, $x = 18$, $y = 7.4$, $z = 8.3$

9 The length of the cylinder is 214 mm, to the nearest millimetre.

10 a $x = 25$ **b** $y = 7$

11 $x = 2.5$

12 The two identical smaller spheres will each have a radius of 15.9 cm, to the nearest mm.

The total surface area of the two small spheres is 6333 cm² compared with 5027 cm² for the larger sphere (nearest cm²).

Hence the two smaller spheres have a greater surface area (by about 26%).

13 The cylinders will be of length 21 cm and 42 cm.

The surface area increases from 5542 cm² (nearest cm²) to 8005 cm² (nearest cm²), an increase of approximately 44%.

14 When the container holds 0.5 litres of liquid the depth of the liquid is 12.4 cm (correct to 1 decimal place).

1 Answers not given here. Check your answers with those of others in your class.

2 $117.40

3

	A & E	Mid	Int/Care
Anne	1	0	1
Jo	1	1	0
Robyn	0	0	1
Rosemary	1	1	1

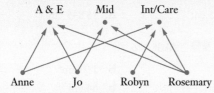

4 a $2 + (3 \times 5)^2 = 227$

b $(2 + 3 \times 5)^2 = 289$

c $(2 + 3) \times 5^2 = 125$

5 $\begin{bmatrix} 12 & 7 \end{bmatrix}$

6 The 450 g box of brekky cereal for $5.90 is the best buy.

7 $5000 invested for 5 years at 8.2% per annum compounded every 6 months. Interest $2472.70.

$5000 invested for 5 years at 8% per annum compounded every 3 months. Interest $2429.74.

$5000 invested for 5 years at 8.5% per annum simple interest. Interest $2125.00

8 a The surface area of the Earth is approximately 5.1×10^8 km^2.

b Approximately 3.6×10^8 km^2 of the Earth's surface is covered by water.

9

Unit cost	Number ordered	Sub total	Less 20% discount	Plus 10% tax
$34.50	17	$586.50	$469.20	$516.12
$8.50	23	$195.50	$156.40	$172.04
$145.50	13	$1891.50	$1513.20	$1664.52
$8.00	56	$448.00	$358.40	$394.24
$1024.00	8	$8192.00	$6553.60	$7208.96

10 a The percentage profit is, to the nearest percent, 33%.

b The item must be sold for at least $31.20.

c The seller purchased the item for $2450.

d The person putting the item into the auction paid $3080 for the item originally.

11 $\begin{bmatrix} 7 & 0 & 0 \\ 0 & 7 & 0 \\ 0 & 0 & 7 \end{bmatrix}$

12 The area of the square is 11 250 cm^2.

13 The area of the rectangle is 360 cm^2.

14 To the nearest whole percent, 36% of the circle is shaded.

15 a The new total surface area is 119% of the original surface area (to nearest percent).

b The new total surface area is 148% of the original surface area (to nearest percent).

Exercise 10A   PAGE 187

1 $960

2 $270

3 15 minutes

4 $270

5 $560

Note: For some boxes a bigger box would need thicker card and hence this would be a volume question with an answer of $1120 (= $28 \times 2^3 \times 5$). However this question does say *using the same thickness and type of card* and so thickness remains unchanged. Hence the question involves considering areas to give the answer of $560 (= $28 \times 2^2 \times 5$).

6 1.125 km^2

7 $560 000

8 86.4 m^3, i.e. about 85 m^3.

9 0.8 cm^2

10 20 cm^2

11 a 3 : 2

b 9 : 4

12 a 4 : 5

b 64 : 125

Exercise 10B   PAGE 191

1 a 12 cm　**b** 30°　**c** 9 : 4

2 a 6.25 cm　**b** 20°　**c** 16 : 25

3 a 7.5 cm　**b** 36 : 25　**c** 216 : 125

4 a 3300 km　**b** 2600 km

c 700 km　**d** 2600 km

e A little over 900 km　**f** Yes

5 a 5.8 m × 6 m　**b** 3.5 m × 4 m

c 3.5 m × 2.7 m　**d** 4.1 m × 1.6 m

6 a Approximately 210 metres.

b Approximately 320 metres.

c Approximately 1250 m^2.

d Approximately 3000 m^2.

Exercise 10C PAGE 196

1 Shapes A and B are similar.

   Lengths in shape A : corresponding lengths in shape B = 2 : 3

2 Shapes A and B are not similar.

3 Shapes A and B are similar.

   Lengths in shape A : corresponding lengths in shape B = 1 : 2

4 Shapes A and B are similar.

   Lengths in shape A : corresponding lengths in shape B = 4 : 1

5 Shapes A and B are not similar.

6 Shapes A and B are similar.

   Lengths in shape A : corresponding lengths in shape B = 2 : 1

7 Shapes A and B are not similar.

8 Shapes A and B are similar.

   Lengths in shape A : corresponding lengths in shape B = 3 : 1

9 No   10 No
11 Yes   12 No
13 No   14 Yes

Exercise 10D PAGE 200

1 $\triangle ABC \sim \triangle DEF$, corresponding angles equal.

   Area $\triangle ABC$: Area $\triangle DEF$ = 1 : 9, $x = 7$, $y = 18$.

2 $\triangle PQR \sim \triangle YZX$, corresponding angles equal.

   Area $\triangle PQR$: Area $\triangle YZX$ = 9 : 4, $x = 10$, $y = 24$.

3 $\triangle DEF \sim \triangle UTS$, two pairs of corresponding sides in same ratio and the included angles equal.

   Area $\triangle DEF$: Area $\triangle UTS$ = 25 : 64, $x = 10.4$.

4 Not similar.

5 $\triangle TUV \sim \triangle MNL$, corresponding sides in same ratio.

   Area $\triangle TUV$: Area $\triangle MNL$ = 16 : 9, $x = 60$, $y = 46$.

6 $\triangle UVW \sim \triangle ZXY$, corresponding angles equal.

   Area $\triangle UVW$: Area $\triangle ZXY$ = 1 : 4, $x = 7$, $y = 12$.

7 $\triangle ABC \sim \triangle EDC$, corresponding angles equal.

   Area $\triangle ABC$: Area $\triangle EDC$ = 9 : 25, $x = 3$, $y = 4.5$.

8 $\triangle PQT \sim \triangle SRT$, corresponding angles equal.

   Area $\triangle PQT$: Area $\triangle SRT$ = 16 : 9, $x = 16$, $y = 15$.

9 $\triangle ABE \sim \triangle ACD$, corresponding angles equal.

   EB = 5 metres.

10 $\triangle PQT \sim \triangle PRS$, corresponding angles equal.

   TS = 10.5 metres.

Miscellaneous exercise ten PAGE 202

1 **a** length   **b** length   **c** area
   d area   **e** area   **f** length
   g volume   **h** area   **i** area
   j area   **k** area   **l** length
   m volume   **n** volume   **o** volume

2 **a** 1.23×10^6 **b** 1.2×10^{-3} **c** 2.5×10^4
   d 2.45×10^{-8} **e** 1.5×10^{10} **f** 3×10^{-5}
   g 7.6×10^1 **h** 1×10^{-1}

3 20 cm, 32 cm and 40 cm.

4 45 bags

5 The 250 g pack of butter costing $4.85 is the better deal.

6 **a** 7.7 km   **b** 4 cm   **c** 0.25 km^2

7 7 : 5

8 $2000 invested for 4 years at 8% per annum compounded six monthly earns more interest by $109.21*.

   (*If you got $309.21 this is the difference in the final amounts, not the difference in the interest.)

9 $AA = \begin{bmatrix} 4 & 0 \\ 1 & 1 \end{bmatrix}$, $AC = \begin{bmatrix} -2 \\ -3 \end{bmatrix}$, $BA = \begin{bmatrix} 5 & -3 \end{bmatrix}$,

   $CB = \begin{bmatrix} -1 & -3 \\ 2 & 6 \end{bmatrix}$, $BC = [5]$.

10 The length of the scale model is 22 cm.

11 Shop B ($1.229 per 100 g), Shop C ($1.326 per 100 g), Shop A ($1.36 per 100 g).

12 **a** 31 km   **b** 20 km^2

13 **a** There will be 1000 smaller cubes each of side length 1 cm.

    b The total surface area of the 1000 small cubes will be 10 times that of the surface area of the initial cube. (6000 cm^2 compared to 600 cm^2.)

14 **a** $1350   **b** $1850   **c** $2350

15 125   16 54 seconds

17 3200 cm^3

18 **a** 2.41 grams   **b** 43.2 kg

19 The top of the ladder will move 66 cm down the wall, to the nearest centimetre.

20 We will assume that Jen's rate of polishing (in area polished per unit of time) stays the same throughout. The time taken will then only depend on the surface area that requires polishing.

The larger sphere is twice the volume of the smaller one.

Hence comparison of length would involve $\sqrt[3]{2}$ and comparison of surface areas would involve $\left(\sqrt[3]{2}\right)^2$. Jen will take 10 minutes $\times \left(\sqrt[3]{2}\right)^2$, i.e. approximately 16 minutes to polish the larger of the two spheres.

ISBN 9780170390194

INDEX